TRANSACTIONS

OF THE

AMERICAN PHILOSOPHICAL SOCIETY

HELD AT PHILADELPHIA
FOR PROMOTING USEFUL KNOWLEDGE

NEW SERIES—VOLUME 40, PART 1
1950

GEOLOGY OF THE COASTAL PLAIN OF NORTH CAROLINA

HORACE G. RICHARDS

*Academy of Natural Sciences of Philadelphia
and University of Pennsylvania*

THE AMERICAN PHILOSOPHICAL SOCIETY
INDEPENDENCE SQUARE
PHILADELPHIA 6

AUGUST, 1950

Reprinted June 1963

GEOLOGY OF THE COASTAL PLAIN OF NORTH CAROLINA

HORACE G. RICHARDS

CONTENTS

INTRODUCTION

SCOPE

The present report deals with the geology of the Coastal Plain of North Carolina. This covers the area east of the "Fall Line" or, approximately, east of a line through Roanoke Rapids, Raleigh, Southern Pines, and Rockingham. The Coastal Plain area cannot be precisely defined because many of the Coastal Plain sediments overlap onto the present Piedmont Plateau. Also there are outliers or remnants of Coastal Plain sediments considerably west of the "Fall Line."

This report covers all the formations of the Coastal Plain from the Cretaceous to the Recent, although relatively less attention is paid to the Cretaceous than to the later sediments. This omission is intentional inasmuch as the deposits and fauna of the Marine Cretaceous have been treated relatively recently (1923) in Volume 5 of the North Carolina Geological and Economic Survey by Dr. L. W. Stephenson.

The major part of the field work for this report was done between January, 1941, and August, 1942. During the latter part of the program, after the entrance of the United States into the war, greater emphasis was paid to the present and potential economic resources of the Coastal Plain. The field work was not carried out continuously during the above mentioned interval, but rather consisted of some fifteen shorter trips interspersed with a study of the fossils collected.

The most extensive report previously published on the Coastal Plain of North Carolina is Volume 3 of the North Carolina Geological and Economic Survey. This was published in 1912 and included contributions by William B. Clark, Benjamin L. Miller, L. W. Stephenson, B. L. Johnson, and Horatio N. Parker. This report has been of inestimable value throughout the present survey and has been drawn upon freely. Subsequent to the writing of Volume 3,[1] considerable progress has been made on unravelling the stratigraphy and geological history of the Coastal Plain of North Carolina and adjacent regions. Therefore, some of the interpretations used in Volume 3 are not tenable in the light of present knowledge. This does not detract from the value of the factual information in that volume. The present report, in addition to recording all new information discovered by the survey, has attempted to bring up to date the data presented in Volume 3.

---

[1] Throughout this manuscript the 1912 report of the North Carolina Geological and Economic Survey will frequently be referred to as "Volume 3."

A previous work by the present writer (Richards, 1936a) on the Pleistocene of the Atlantic Coastal Plain has also been freely drawn upon in this report.

A preliminary manuscript was submitted to the North Carolina State Museum in September, 1942. Since that time it has been possible to do additional field work in North Carolina, and an attempt has been made to incorporate as much additional material as possible into this report.

This manuscript has been released for publication by the North Carolina State Museum and the Division of Mineral Resources of the Department of Conservation and Development of the State of North Carolina.

ACKNOWLEDGMENTS

The present survey was part of a joint project sponsored by the Academy of Natural Sciences of Philadelphia and the North Carolina State Museum of Raleigh, North Carolina. A large part of the field work was made possible by a grant from the Johnson Fund of the American Philosophical Society, to which sincere thanks are due.

The writer is indebted to the Academy of Natural Sciences, in particular to Dr. B. F. Howell, Curator, for making possible the Academy's part of the cooperation and for much interest throughout the course of the work.

He is also indebted to Mr. Harry T. Davis, Director of the North Carolina State Museum, for arranging the State's part in the cooperative program. Mr. Davis carried the writer on a large part of the field work and offered the benefit of his wide experience in the Coastal Plain of North Carolina. He also gave numerous suggestions during the preparation of the manuscript.

The Division of Mineral Resources, North Carolina Department of Conservation and Development, through Dr. J. L. Stuckey, State Geologist, fully cooperated by making available all pertinent information. Furthermore, Mr. M. J. Mundorff, of the Ground Water Division of the United States Geological Survey, took part in several field conferences and supplied valuable information in the nature of locality records, well logs, etc.

Thanks are due to the various County Agents of Eastern North Carolina who cooperated by supplying information in regard to their respective counties. The late L. W. Anderson, County Agent of Perquimans County, reported the important new Pleistocene fossil locality near Nicanor, North Carolina.

Thanks are also due Dr. B. W. Wells, North Carolina State College; Dr. John Parker, North Carolina State College, Raleigh; W. R. Robeson, Tar Heel; J. H. Cannaday, Kinston; J. W. Butler, Acme; J. T. Smith, Cameron; and C. L. Hardy, Maury. Other names will be found throughout the text in reference to the respective localities. Mr. Allen L. Midyette, Jr., served as field assistant throughout much of the work and also helped with the photography.

At the invitation of Dr. William S. Pike, of the Shell Oil Company, Incorporated, the writer participated in several field conferences held in the Coastal Plain of North Carolina during 1945 and 1946, and he wishes to express his appreciation for this courtesy. He was also invited to visit the drilling operations of Standard Oil Company of New Jersey in Dare County in 1946 and 1947, and is indebted to Mr. K. D. White and Walter B. Spangler for this invitation and for supplying him with information and samples.

Dr. C. Wythe Cooke, Mr. M. J. Mundorff, and Mr. Harry LeGrand, of the United States Geological Survey, and Dr. Richard F. Flint of Yale University kindly read certain portions of this manuscript and gave the benefit of their constructive criticism.

Finally, thanks are due to Miss Helen Shoemaker and Mr. Albert Burkhardt for assistance in sorting the fossils and in preparing the plates for publication.

STRATIGRAPHY OF THE NORTH CAROLINA COASTAL PLAIN

Table 1 shows the present interpretation of the stratigraphy of North Carolina. This is presented in abbreviated form in order to serve as background for the detailed sections to follow.

TABLE 1

COASTAL PLAIN FORMATIONS OF NORTH CAROLINA

PLEISTOCENE
- PAMLICO — marine; last interglacial
- HORRY — clay; low sea level; glacial
- TALBOT
- PENHOLOWAY
- WICOMICO — "Higher Terraces" — no marine fossils
- SUNDERLAND
- COHARIE

PLIOCENE
- High Level Gravels ("Lafayette")
- CROATAN near Neuse River — may be contemporaneous
- WACCAMAW near Cape Fear River

MIOCENE
- DUPLIN (south of Neuse River)
- YORKTOWN (north of Neuse River)
  - Zone 2
  - Zone 1 ("Murfreesboro Stage")
- ST. MARY'S (possibly buried in northeastern North Carolina)
- CALVERT (buried near Cape Hatteras)
- TRENT (mostly in Onslow County)

OLIGOCENE — —unnamed— (buried in Onslow County)

EOCENE
- CASTLE HAYNE (Jackson)
- Lower and Middle Eocene (in wells and locally at western limit of Coastal Plain)

UPPER CRETACEOUS
- PEEDEE
- BLACK CREEK
- TUSCALOOSA

LOWER CRETACEOUS
- Possibly part of TUSCALOOSA
- —unnamed— (in deep wells)

TABLE 2

COMPARISON OF THE PRESENT INTERPRETATION OF THE
STRATIGRAPHY OF THE COASTAL PLAIN OF NORTH
CAROLINA WITH THAT OF THE PREVIOUS
VOLUME (1912)

	Volume 3 (1912)	Present Interpretation (1950)
PLEISTOCENE	PAMLICO ——— CHOWAN WICOMICO SUNDERLAND COHARIE	PAMLICO HORRY CLAY TALBOT PENHOLOWAY WICOMICO SUNDERLAND COHARIE
PLIOCENE	LAFAYETTE WACCAMAW	"High Level Gravels" CROATAN WACCAMAW
MIOCENE	DUPLIN YORKTOWN ST. MARY'S ——— ———	DUPLIN Zone 2 of YORKTOWN Zone 1 of YORKTOWN ("Murfreesboro") ST. MARY'S (possibly buried) CALVERT (buried) TRENT
OLIGOCENE	———	—unnamed— (buried)
EOCENE	CASTLE HAYNE TRENT	beneath Trent = Lower Miocene Lower and Middle Eocene
UPPER CRETACEOUS	PEEDEE BLACK CREEK (including Snow Hill)	PEEDEE BLACK CREEK (including Snow Hill)
LOWER CRETACEOUS	PATUXENT	TUSCALOOSA (Basal Upper Cretaceous) Part of TUSCALOOSA may be Lower Cretaceous Marine LOWER CRETACEOUS in wells near Hatteras

Table 2 shows a comparison between the present interpretation and that used in the previous volume on the North Carolina Coastal Plain (N. C. Geol. and Econ. Surv. Vol. 3, 1912).

The present interpretation is based upon a careful study of previous work and has also incorporated a few minor changes suggested during the course of the present survey.

For correlations of the formations of North Carolina with those from other states, see table on page 54.

A new geological map of the Coastal Plain of North Carolina has recently been published by North Carolina Department of Conservation and Development, and is included in a report on marls by Dr. Willard Berry (1947=1949).

LOWER CRETACEOUS

The deposits referred to the Patuxent formation of Lower Cretaceous age in Volume 3 are now generally regarded as being largely or entirely of Upper Cretaceous age, and are referred to the Tuscaloosa formation.

Recently the presence of Lower Cretaceous sediments underlying Eastern North Carolina has been demonstrated by a study of samples from several oil tests drilled in that region (Swain, 1947).

In the well drilled by the Standard Oil Company of New Jersey at Cape Hatteras, Lower Cretaceous sediments were reported between the depths of 6,700 and 9,878 feet, while in the same company's well in Pamlico Sound these beds were encountered from 5,495 feet to the bottom of the well at 6,410, and they presumably extended still deeper. These beds are correlated with the Washita, Fredericksburg, Trinity, and Coahuila (?) formations of the Gulf Coast by Swain (1947) who has made a study of the microfossils. Similar Lower Cretaceous microfossils have been found in the deep wells in Cartaret County, but the exact correlations have not yet been worked out.*

UPPER CRETACEOUS

TUSCALOOSA FORMATION

NAME

This formation name was proposed by Smith and Johnson in 1887 from Tuscaloosa, Alabama. In North Carolina the formation now referred to the Tuscaloosa was named the "Cape Fear" by Stephenson in 1907. This he tentatively correlated with the Patuxent of Maryland of Lower Cretaceous age. Later, in 1912, he substituted the name Patuxent.

In South Carolina the local names "Hamburg" and "Middendorff" from towns in Aiken and Chesterfield Counties were applied by Sloan (1904) to clays and sand supposed to be of Lower Cretaceous age. However, Berry (1914) from a study of the plants showed that the "Middendorff" beds were of Upper Cretaceous age. From later work along the Coastal Plain, Cooke (1936) showed that the "Middendorff" beds should be correlated with the Tuscaloosa formation of Alabama which is of basal Upper Cretaceous age.

Although little additional field work has been done on the "Patuxent" beds of North Carolina, it has been shown by Cooke (1936) and Stephenson (1942) that the Tuscaloosa beds of South Carolina extend into North Carolina and are equivalent to at least part of the so-called "Patuxent." It is possible that part of the "Patuxent," as mapped in northeastern North Carolina, really is older than the Tuscaloosa and actually belongs to the Lower Cretaceous. However, in the absence of fossils, this cannot be definitely demonstrated. Mun-

---

* In a paper presented before the American Association of Petroleum Geologists in April 1950 at Chicago, Swain, on the basis of the ostracod fauna, suggested that the lower 1,400 feet of the Cape Hatteras well (Esso No. 1) may be of Jurassic age.

dorff (1946) uses the term "undifferentiated Cretaceous" for beds of this age in the Halifax area.†

EXTENT

The Tuscaloosa in North Carolina occurs in a belt beginning at the Roanoke River at the boundary between the eastern halves of Northampton and Halifax Counties, passes to the southwest through the following counties: Edgecombe, northern Pitt, Wilson (south of the town of Wilson), northern Greene, north half of Wayne, Johnston (south of Smithfield), northern Sampson, southern Harnett, almost all of Cumberland, southern Moore, northern Robeson, all of Scotland, and southern Richmond to the South Carolina line. North of Wayne County the occurrences of the Tuscaloosa formation represent erosional inliers rather than extensive outcrops.

It extends southward into South Carolina where it has been mapped by Cooke (1936). In Virginia the Tuscaloosa is known only from wells near Norfolk and Franklin.

UNDERLYING FORMATIONS

The Tuscaloosa rests on Paleozoic or Pre-Paleozoic crystalline rocks, except in a few places where it lies on the Triassic. In wells in eastern North Carolina, the Tuscaloosa formation grades downward conformably into the Lower Cretaceous.

OVERLYING FORMATIONS

The Tuscaloosa formation is always overlain unconformably by the Black Creek or some later formation. Along the Cape Fear River it is overlain by the Black Creek. According to Volume 3, this contact is best seen between milepost 100 and 101 (above Church Landing). A similar contact was observed along Highway 701 between Clinton and Smithfield at Stone Creek crossing in Johnston County. Locally, near Spout Springs in Harnett County, the Tuscaloosa is overlain by the Eocene. From the southwest edge of Wilson County northeast to the Roanoke River, the Tuscaloosa is overlain by overlapping beds of Miocene or Pleistocene terrace deposits. A Miocene contact is well exposed at Shiloh Mills on Tar River, Edgecombe County, and at Palmyra Bluff on the Roanoke River. It is overlain by the "Brandywine" in the "Sand Hills" of Richmond, Scotland, Cumberland, Hoke, Moore, and Harnett Counties.

LITHOLOGY

The Tuscaloosa formation is composed of sands and clays. It is mostly arkosic and often contains pure white kaolin grains. In some places in Harnett and Hoke County, the kaolin is remarkably pure and is possibly of some economic value. Mica grains are also present, particularly where the formation is near a contact with the crystalline rocks. Iron sulphide (marcasite) is occasionally present in the clay beds, for instance at milepost 105 on the Neuse River. A small amount of lignite is present on Contentnea Creek and Tar River. In the subsurface, the Tuscaloosa frequently consists of reddish clay. Marine sands and clays are known in deep wells in extreme eastern North Carolina.

THICKNESS

The Tuscaloosa formation is variable in thickness, possibly indicating an uneven surface of the basement rock on which it was deposited. It appears to dip toward the coast, at about the rate of 14 feet per mile. East of a line through Beaufort, the dip increases greatly to 122 feet per mile between Beaufort and Cape Hatteras.

The following thicknesses are reported from studies of well samples (after Stephenson, 1912: 87, and Richards, 1945: 910–916; 1948 and unpublished data):

Goldsboro	80 feet
Wilson	90 feet
Tarboro	263 feet
Maxton	351 feet
Scotland Neck	275 feet
Wilmington	22 feet
Havelock	193 feet
Fayetteville	255 feet
Merrimon	790 feet
Hatteras	1,784 feet

CORRELATION

As stated above, this formation has been correlated with the Tuscaloosa beds which outcrop along the inner margin of the Coastal Plain from North Carolina to Mississippi. From a study of the fossil plants found at various places in South Carolina, Alabama, and elsewhere, a basal Upper Cretaceous age (Cenomanian) has been suggested (Berry, 1914, etc.). Except in deep wells and a few obscure casts in Alabama, no marine fossils have been found. Farther south the Tuscaloosa is thought to be correlated with the Woodbine formation of Texas with a fairly rich marine fauna. To the north the Tuscaloosa is now correlated with the Raritan formation of Maryland, Delaware, and New Jersey. This correlation is largely based on the fossil plants as reported by Berry (1910, 1920). Recently, a marine fauna has been described from the Raritan at Sayreville, New Jersey (Richards, 1943c), giving further evidence for a correlation between that formation and the Tuscaloosa. Still farther north, Stephenson (1936) recently described some marine mollusks dredged off Banquereau, Nova Scotia, which he correlated with the Raritan and Tuscaloosa. Therefore, while most of the Tuscaloosa is apparently of non-marine origin, there are phases of the formation which are of marine origin. Part of the Tuscaloosa, as mapped in northeastern North Carolina, may be of Lower Cretaceous age, equivalent to the Patuxent of Virginia and Maryland.

---

† Recent papers by Spangler and Peterson (1950) and Spangler (1950) propose a correlation of the Tuscaloosa formation of North Carolina and the "Patuxent" formation of Virginia and that they date from both Early and Late (Lower and Upper) Cretaceous time.

DETAILED SECTIONS

Since the Cretaceous formations of North Carolina are usually found exposed only along stream banks, their distribution will be discussed first by river; then the localities away from the main rivers will be discussed under the respective counties.

Cape Fear River (Harnett and Cumberland Counties)

The Tuscaloosa is exposed from 1½ miles above the mouth of Little River, Harnett County, to Devanes Ferry, 17 miles below Fayetteville, Cumberland County. Between mileposts 100 and 101 it is overlain by the Black Creek formation which is exposed along the Cape Fear River below Devanes Ferry (fig. 4).

Neuse River (Johnston and Wayne Counties)

From 1 mile above Cox's Bridge, Johnston County, to the Atlantic Coast Line Railroad bridge southwest of Goldsboro, Wayne County.

Contentnea Creek (Wilson and Greene Counties)

From 1 mile above Woodard's Bridge (8 miles SSE Wilson) to Fools Bridge near Contentnea. A good section is seen ½ mile above Speights Bridge. The following section is from Volume 3, page 100:

	Feet
Concealed by vegetation	12
Pleistocene	
Mostly concealed, but consisting of sand with a band of gravel at the base	12–15
Miocene	
Sand and clay	9
Cretaceous (Tuscaloosa)	
Very compact drab, sandy micaceous clay	4
Very compact gray, micaceous, argillaceous arkosic sand	6

Tar River (Edgecombe and Pitt Counties)

From 4 miles below Dunbar Bridge, Edgecombe County, to Parkers Landing in Pitt County. It is usually overlain by Miocene or Pleistocene, although at Parker Landing it is overlain by a thin deposit of Black Creek. A good section showing the Tuscaloosa along the Tar River can be seen at Shiloh Mills, 2 miles above Tarboro (fig. 6).

Roanoke River (Halifax County)

From 5 miles below Halifax to Palmyra, Halifax County. Here the Tuscaloosa is overlain by Miocene or Pleistocene. Near Halifax the Miocene rests directly on the bed rock. For section at Palmyra Landing, see page 24.

Johnston County

Road cut east of Stone Creek on Highway 701 between Clinton and Smithfield. The following section was observed in May, 1942:

	Feet
Pebbly sand ....Coharie (Pleistocene)	0–2
Black clay and lignitized wood..Black Creek	7
Gray sandy clay ....Tuscaloosa	1

(Exposed at bottom of ditch)

Volume 3 records an occurrence of the Tuscaloosa along the Atlantic Coast Line Railroad at Four Oaks. It is here overlain by the Pleistocene.

Cumberland County

Tuscaloosa clay is well exposed at many places in Cumberland County. Along the Fort Bragg Boulevard, 2 miles northwest of Fayetteville, typical mottled sandy clay (Tuscaloosa) is overlain by a black carbonaceous clay (Black Creek) which in turn is overlain by pebbly sand of the Coharie formation.

A similar relationship between the Tuscaloosa and the Black Creek was recorded by Stephenson (Vol. 3: 110) at Hope Mills.

Hoke County

The Tuscaloosa formation covers all of Hoke County. Clay deposits, now used for road construction, occur at various places in Fort Bragg Military Reservation. The following pits were carefully examined and found to contain typical arkosic sand with lenses of white clay (kaolin) and some mica: (a) clay pit on Plank Road, Fort Bragg Military Reservation; (b) clay pit on King Road, Fort Bragg Military Reservation.

A careful search was made for bauxite at these localities, but without success.

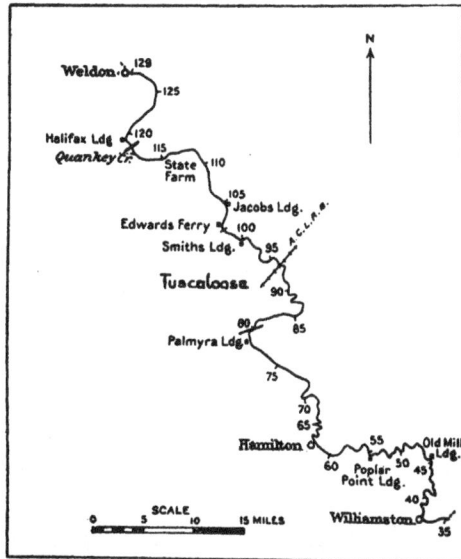

Fig. 1. Sketch map of Roanoke River showing outcropping formations (from N. C. Geol. and Econ. Surv., Vol. 3).

Near the Sanatorium, typical Tuscaloosa arkosic sand is overlain by iron conglomerate of Eocene age.

Harnett County

Spout Springs.—Typical fine Tuscaloosa arkosic sandy clay is exposed along the railroad cut near Spout Springs station (fig. 14).

Chalk Hill.—Shallow excavations on the top of a low hill, 4 miles southwest of Olivia revealed numerous clay balls of unusually pure white kaolin. These were apparently reworked from the Tuscaloosa formation.

Moore County

Southern Pines.—Arkosic sandy clay with small patches of white kaolin can be seen along the railroad cut (Seaboard Air Line Railroad) just north of Southern Pines (fig. 13).

Aberdeen.—There is an extensive sand pit on highway No. 211 just west of Aberdeen. Here the Tuscaloosa sand is overlain by a thin mantle of Brandywine (?) gravel. This pit is now being worked.

Richmond County

Clay, presumably of the Tuscaloosa formation, has been noted near Rockingham and Hamlet, but no detailed studies have been made in this region.

Scotland County

Similar sandy clay has been noted near Laurinburg and elsewhere in Scotland County. This is probably part of the Tuscaloosa formation.

MARINE TUSCALOOSA FOSSILS IN WELLS

Recent work has demonstrated the presence of a marine phase of the Tuscaloosa formation in deep wells under eastern North Carolina. This discovery suggests that the non-marine deposits of the outcropping Tuscaloosa formation grade downward and eastward into a marine phase. A marine phase of the Tuscaloosa has been recognized in the following wells:

		Depth
Kartson-Laughton	Near Morehead City, Cartaret County	2,263–2,900
Carolina Petroleum Company wells	Near Merrimon, Cartaret County	2,210–3,000 [2]
Esso No. 1	Cape Hatteras, Dare County	4,800–6,584
Esso No. 2	Pamlico Sound, Dare County	3,890–5,495

A fauna of microfossils has been obtained from some of these wells and a preliminary report on those from the Esso wells has been published (Swain, 1947). A more complete report is in preparation and will be published by the United States Geological Survey (Swain). Among the macrofossils, *Exogyra woolmani* Richards

[2] Generalized from several wells near Merrimon.

and *Hamulus protoonyx* Richards are characteristic forms (see Richards, 1948a).

BLACK CREEK FORMATION

NAME

The name Black Creek Shale was originally proposed by Sloan (1907) for beds exposed along Black Creek in Florence and Darlington Counties, South Carolina. In 1907 Stephenson applied the term "Bladen" formation to beds of Upper Cretaceous age in North Carolina. Later (Vol. 3, 1912), observing the priority of the term Black Creek, he abandoned the name "Bladen."

In 1923 Stephenson used the term Snow Hill calcareous member for the upper part of the Black Creek formation consisting of laminated sands and clays interstratified with layers or lenses of more or less calcareous greensand and marine clay, often with many fossils. The type locality was given as Snow Hill, Green County, North Carolina, and other typical exposures were given as: Blue Banks Landing on Tar River, Pitt County, North Carolina; numerous localities on Black River in Sampson County, North Carolina; and at Hodge's old mill site in Marlton County, South Carolina.

EXTENT

The Black Creek formation outcrops in a belt southeast of that described for the Tuscaloosa. In the Cape Fear River region it has a width of some 30 miles or more, but it narrows toward the north and in the Tar River Region is only about 8 miles wide. The Black Creek outcrops are found in the following counties: northern Pitt, northern Greene, Wayne, northwest Duplin, Sampson, southern Cumberland, Bladen, and Robeson.

The Black Creek formation usually rests unconformably on the Tuscaloosa and is overlain conformably by the Peedee formation. It extends eastward beneath the Peedee as shown by wells at Wilmington and elsewhere. Contacts between the Black Creek and Peedee can be seen along the Cape Fear River at Donohue Creek Landing.

Farther inland than the outcrops of the Peedee formation the Black Creek is frequently overlain by Pleistocene terrace deposits and is only exposed in stream cuts. A typical exposure of the Black Creek overlain by the Coharie terrace deposit was seen near Stone Creek in Johnston County (see section on page 37). Along the Neuse River below Kinston, the Black Creek is overlain by the Eocene (Castle Hayne formation) and near Elizabethtown it is covered locally by shell marl of Miocene (Duplin) or Pliocene age (Waccamaw).

LITHOLOGY

The Black Creek formation is largely composed of thinly laminated sand and clay with a slight glauconitic content. Lignite is common and iron sulphide (marcasite and pyrite) is known locally. The upper part of the formation is more highly glauconitic and contains in places a rich molluscan fauna. This has been designated by Stephenson as the Snow Hill member, named from Snow Hill, Greene County, North Carolina.

DIP

"Slight," less than 20 feet per mile, increasing toward the coast.

THICKNESS

The thickness is not easy to determine because contacts are not always sharp. The formation is about 300 feet thick in wells at Wilmington and Morehead City, but the increased dip northeast of Morehead has caused a thickness to some 600 feet in the Hatteras well. However, the older part of the Black Creek (Magothy or Eutaw) is not included in these figures.

DETAILED SECTIONS

Cape Fear River (Cumberland and Bladen Counties)

The Black Creek formation outcrops along the Cape Fear River from a point 101 miles by river above Wilmington (near Church Landing, Cumberland County) to Jessups Landing, 56 miles above Wilmington in Bladen County. Detailed sections of many localities are given by Stephenson (Vol. 3, 1912; Vol. 5, 1923). Subsequent to Stephenson's field work, various locks have been constructed along the river with the result that many of the localities are no longer accessible, having been washed away or covered.

The "feather edge" of the formation has been observed near Hope Mills, Cumberland County. Here it rests on the Tuscaloosa formation. Fossil plants occur in the Black Creek formation at various places between Fayetteville and Elizabethtown. Some of these have been listed in Volume 3, while others are discussed by Berry (1920) (fig. 4).

Petrified wood described by Boeshore and Gray (1936) as *Torreya antiqua* has been found along a small branch (tributary of the Cape Fear) about 5 miles south of Fayetteville and 100 feet west of Fayetteville-Elizabethtown highway (No. 87). This locality was discovered by Mr. H. A. Rankone, of Fayetteville.

Among the other localities where fossil plants have been discovered are the following:

Rockfish Creek, 7 miles south of Fayetteville near highway 87 (Rankone)
Prospect Hall Bluff, Cape Fear River (near milepost 93) (Berry)
Mouth Harrisons Creek (milepost 83) (Berry)
Court House Landing (milepost 77) (Berry)
Mines Creek, 20 miles below Fayetteville (Rankone)
Roadcut near bridge, Elizabethtown (Berry)

No marine fossils have been found in the Black Creek formation between Fayetteville and Elizabethtown. According to Stephenson (1923: 38), these deposits can probably be correlated with the *Exogyra upatoiensis*

zone as shown in the Charleston well. This is stratigraphically lower and older than the Snow Hill member and may be equivalent to the Magothy formation of New Jersey.

Phoebus Landing.—Many years ago a number of poorly preserved dinosaur bones and other fossil remains were obtained at Phoebus Landing on the right bank of the Cape Fear River near milepost 68, some 4 miles below Elizabethtown, Bladen County. These have been identified by C. W. Gilmore as:

Dinosauria
    Hypsibema crassicaude Cope
    Trachodon tripos? Cope
    Zatomis sp.?
Crocodylidae
    Thecachampsa rugosa Emmons
    Polydectes biturgidus Cope
Testudinata
    Taphrosphyds dares Hay
    Amyda sp.

The locality is now (1942) completely covered with silt and debris and impossible to find.

At *Walkers Bluff,* near milepost 60, 13 miles below Elizabethtown by water or 8 miles by road, on the property of Mr. Munrow, the Cretaceous is overlain by the Waccamaw, Pliocene. For detailed section see Stephenson (Vol. 3: 121).

A few Cretaceous fossils were obtained during the present survey, although no attempt was made to make a thorough collection from this formation.

At *Jessups Landing,* left bank, milepost 56, marine Black Creek fossils have been recorded by Stephenson.

Below this point questionable Black Creek is reported at Whitehall Landing (near milepost 53) and Deep Water Point (near milepost 51). From Donahue Creek Landing (50½ above Wilmington), down stream the Peedee formation forms the banks of the Cape Fear River.

Black River (Sampson County)

Black Creek deposits occur along Black River between Bradshaws Landing (74¾ miles above Wilmington) and Horrells Landing (48¾ miles above Wilmington) and possibly slightly above the former locality. The following fossil localities have been recorded above Wilmington by Stephenson:

Bradshaws Landing (74¾ miles) plants
Sykes Landing (74 miles) plants
Big Bend (73¾ miles) plants
Mossy Log Landing (71½ miles) mollusks
Milepost 69 mollusks
Bryan Newkirks marl hole (66 miles) mollusks
Milepost 64 mollusks
Corbitts Landing (63¼ miles) mollusks
62½ miles above Wilmington mollusks
Kerrs Cove (62¼ miles) mollusks

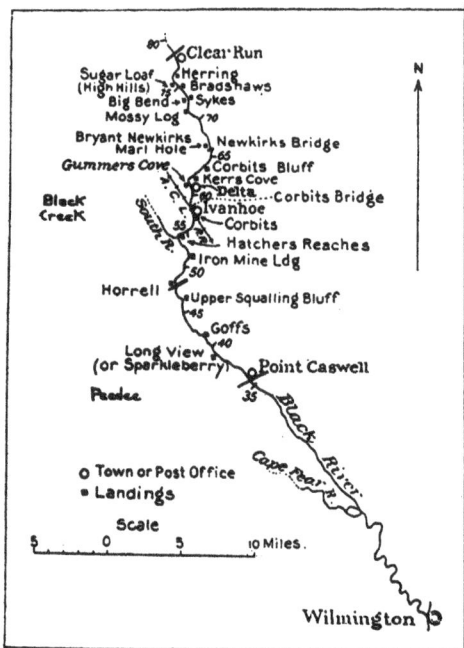

FIG. 2. Sketch map of Black River showing outcropping
formations (from N. C. Geol. and Econ. Surv., Vol. 3).

A.C.L. RR. bridge (58 miles) plants
Corbitts (Old Union) bridge (57¾ miles) plants
Near Ivanhoe (about 58 miles) plants and mollusks
56¾ miles above Wilmington plants
Hatcher Reaches (54½ miles) mollusks
Iron Mine Landing (51 miles) mollusks
Horrell Landing (48¾ miles) plants

Below this point, the Black Creek is buried beneath
the Peedee Formation which crops out along Black
River between Goff Landing (41¼ miles) and Point
Caswell (36 miles).
No attempt was made to check these various localities
during the present survey, although a collection of typical Black Creek fossils was obtained from Kerrs Cove.

Neuse River (Wayne and Lenoir Counties)

Exposures of the Black Creek formation (Snow Hill
stage) occur in the bluffs of the Neuse River from
Blackmans Bluff (117½ miles above New Bern), to
Whiteley Creek Landing (60 miles above New Bern
and 10 miles above Kinston, Lenoir County) (fig. 5).
The following are the better known localities:

Blackmans Bluff, right bank—fossil plants
Near Arringtons Bridge, milepost 92—plants, casts
of shells
87½ miles above New Bern—plants and casts of shells

79¼ miles above New Bern, right bank—shells
Seven Springs—Recently an unusually perfect speci-
men of *Placenticeras placenta* (DeKay) was ob-
tained at this locality; the specimen is now in the
North Carolina State Museum
Auger Hole Landing, left bank, milepost 73—many
mollusks
Whiteley Creek Landing—many mollusks, right bank

The best collections of mollusks came from Auger
Hole Landing and Whiteley Creek Landing. These
fossils are very similar to those from Snow Hill. No
attempt was made during the present survey to collect
from any of these Neuse River localities except at Seven
Springs.
Farther down the Neuse River, between Williams
Landing (52 miles above New Bern) and the mouth of
Contentnea Creek, the banks of the river are composed
of the Peedee formation which overlies the Black Creek.

Contentnea Creek (Greene County)

The Black Creek formation occurs along Contentnea
Creek from a point 2 miles above Speights Bridge (20
miles above Snow Hill) to a point about 6 miles below
Snow Hill. For the first few miles upstream the formation rests unconformably on the Tuscaloosa. Near
Snow Hill the Tuscaloosa is deeply buried (fig. 5).
Snow Hill.—The classic North Carolina Cretaceous
locality is at Snow Hill. This is the type locality of the
Snow Hill stage of Stephenson. The best material was
collected many years ago by Kerr, Stanton, Gabb, Conrad, and others. Stephenson was not able to obtain any
fossils along Contentnea Creek near Snow Hill, although

in a ravine near an old schoolhouse, in the scarp bordering
a swamp to the west of town, a small fossiliferous exposure
was found in which were collected a large number of fossils
including the larger part of the species described by Conrad
in Kerr's report. . . . The matrix here consists of dark
green, glauconitic, argillaceous sand or sandy clay. The
fossil layer is perhaps 10 to 15 feet above medium low water
level in Contentnea Creek.

During the present survey no fossils were obtained
from the banks of Contentnea Creek and only a few small
fragments could be found in the above-mentioned ravine.
Stephenson recorded several exposures of the Snow
Hill stage between Snow Hill and a point about 6 miles
downstream, but none were found during the present
survey. Below that point to Grifton, Pitt County, the
banks of the Contentnea Creek consisted of the overlying Peedee formation.
Maury.—The Snow Hill stage of the Black Creek
underlies the Miocene Yorktown in the vicinity of
Maury, Greene County. Among the specimens dug
from the C. L. Hardy pits, 2 miles east of Maury, were
a few Cretaceous shells.

Tar River (Pitt County)

The banks of the Tar River contain beds of the Black
Creek formation (Snow Hill stage) interruptedly from

near Parkers Landing, 12 miles above Greenville to Randolph Landing, 6 miles above Greenville. The two best localities are as follows (fig. 6):

One-eighth Mile Below Parkers Landing, Left Bank Tar River.—Here the Black Creek rests on the Tuscaloosa, and contains an abundance of lignite, iron sulphide concretions, and occasional pieces of amber. Nine species of fossil plants have been identified by Berry.

Blue Banks Landing, Right Bank Tar River, 7 Miles Above Greenville.—This locality has yielded a large collection of fossil mollusks, although at the present time, only a few poor specimens could be obtained. This material has been carefully studied by Stephenson (1923).

Johnston County

Stone Creek.—Considerable fossil wood was found in the Black Creek formation at a road cut along highway 701 at Stone Creek Bridge. Here the Black Creek overlies the Tuscaloosa. A similar relationship was reported by Stephenson at Bentonville (Vol. 3: 131) (fig. 15, 16).

New Hanover County

Wilmington.—The buried coastward extension of the Black Creek has been recognized in the well boring for the Clarendon waterworks plant at Wilmington where the thickness amounts to 367 feet, extending from a depth of 720 feet to 1,087 feet below the surface overlying the Tuscaloosa. It is overlain by the Peedee (Stephenson, Vol. 3: 144; Richards, 1945: 916).

Fort Caswell.—A similar well here shows the Black Creek to have a thickness of 400 feet extending from the base of the Peedee at 1,140 feet to Tuscaloosa formation at 1,455 feet (Stephenson, Vol. 3: 144; Richards, 1945: 916).

CORRELATION

The Black Creek formation is apparently equivalent to the Selma chalk of Alabama and Mississippi and the Taylor formation of Texas. To the north it is equivalent to the Matawan group and the Magothy formation of Maryland, Delaware, and New Jersey. In New Jersey, the Matawan group has been divided into five distinct formations (Wenonah, Marshalltown, Englishtown, Woodbury, and Merchantville). However, such subdivisions cannot be recognized in North Carolina. The formations of the Matawan group constitute the *Exogyra ponderosa* zone of Stephenson as characterized by the presence of this large pelecypod.

The older portions of the Black Creek formation, as exposed along the Cape Fear River above Elizabethtown, are non-marine and are characterized by the presence of numerous fossil plants. Similar fossil plants have been found in phases of the Black Creek in South Carolina (near Darlington) and from the Magothy formation of New Jersey (equivalent to the lower part of the Black Creek). It is probably also equivalent to the Eutaw of Alabama. Further evidence for this correlation has been afforded by the finding of Eutaw fossils in the wells at Hatteras, Merrimon, etc.

According to Stephenson (1923: 38) the continental phase of the Black Creek may be correlated with the *Exogyra upatoiensis* zone from the Charleston, South Carolina well. This is stratigraphically lower than the Snow Hill stage.

The rest of the Black Creek formation, including the Snow Hill stage, is definitely of marine origin and is regarded as part of the *Exogyra ponderosa* zone of Stephenson. This fauna has been carefully studied by Stephenson (1923).

TABLE 3

CORRELATION BETWEEN THE BLACK CREEK FORMATION AND CONTEMPORANEOUS FORMATIONS IN NEW JERSEY

NORTH CAROLINA			NEW JERSEY
BLACK CREEK	*Exogyra ponderosa* zone	Snow Hill Stage	Wenonah
			Marshalltown Englishtown Woodbury Merchantville
	E. upatoiensis zone		Magothy

FOSSILS FROM BLACK CREEK FORMATION

The fossils from the Black Creek formation have been described and figured by Stephenson (1923). The best localities for these fossils were at Snow Hill and Blue Banks Landing. Unfortunately, conditions along many of the rivers in eastern North Carolina have changed considerably since these localities were described, and it is now very difficult to find any good collecting localities from the Black Creek formation in North Carolina. A few fossils are occasionally found in excavations or wells, but the number of species is very small as compared with the long lists in Stephenson's book. A few of the more characteristic Black Creek fossils are as follows: *Cucullaea antrosa* Morton; *Exogyra ponderosa* Roemer; *Anomia argentaria* Morton; *Lucina parva* Stephenson; *Placenticeras placenta* DeKay.

PEEDEE FORMATION

NAME

This formation was first described under the name Peedee bed by Ruffin in 1843 (p. 27) from the Peedee River in South Carolina. The same formation was later described by Sloan (1907) under the name "Burches Ferry," from a settlement of that name on the Peedee River in South Carolina. Stephenson, also in 1907, used the term Ripley for these deposits in North Carolina because of the similarity with the Gulf Coast deposits of that name. However, after further work, Stephenson concluded (1912: 145; 1942) that the Ripley and Peedee were not exact equivalents.

EXTENT

In North Carolina the Peedee outcrops in a belt east and southeast of the Black Creek formation. It is widest in the Cape Fear River region (about 37 miles) and extends north-northeast to Greenville. Outcrops occur in the following counties: Pitt, Greene, Lenoir, Duplin, Pender, Bladen, New Hanover, Columbus, and Brunswick.

UNDERLYING FORMATIONS

It conformably overlies the Black Creek formation.

OVERLYING FORMATIONS

It is overlain unconformably by Tertiary deposits. From Wilmington northward to the Neuse River, the Eocene occupies basin-like depressions in the undulating surface of the Peedee. Thin discontinuous patches of Miocene, Pliocene, and Pleistocene locally rest unconformably on the Peedee, or on the intermediate Eocene.

LITHOLOGY

It is composed largely of dark-green or gray finely micaceous, more or less glauconitic and argillaceous sand, many layers being calcareous or impure limestone. Irregular concretions of calcium carbonate occur in places. Dark marine clays are interstratified with sand beds. The glauconite is considerably less concentrated than the Cretaceous deposits of New Jersey.

THICKNESS

Fort Caswell well	886 feet
Wilmington well	720 feet
Hatteras well	420 feet

DETAILED SECTIONS

Cape Fear River (Bladen, Columbus, and Pender Counties)

The Peedee formation outcrops along the Cape Fear River from Donahue Creek Landing (50⅓ miles above Wilmington) to near the mouth of Black River (15½ miles above Wilmington). The following localities are listed by Stephenson, all above Wilmington:

Donohue Creek Landing (50⅓ miles)
Robinsons Landing (49½ miles)
Kelleys Cove (46 miles)
Indian Wells Landing (41 miles)
Neils Eddy Landing (28 miles)
Daniels Landing (40 miles)
Kings Bluff (38½ miles)
Black Rock Landing (37 miles)
Huddlers Landing (30½ miles)
Bryants Landing (27 miles)
Magnolia Landing (15½ miles)

Shells are reported from all these places, although the number of species is not great. The only place visited during the present survey was Neils Eddy Landing, where Peedee clay containing specimens of *Ostrea sub-*

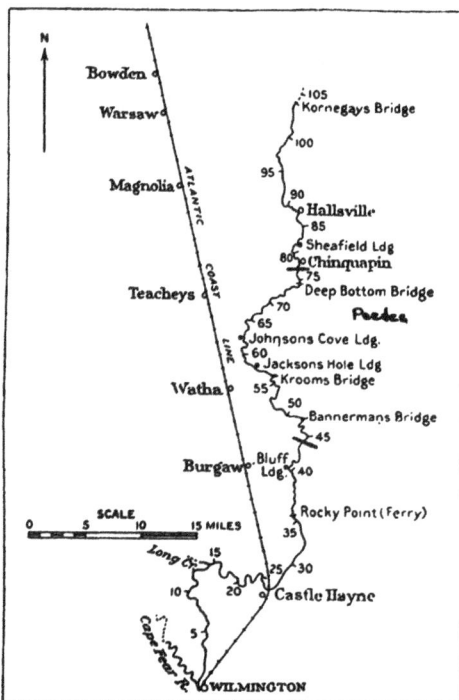

FIG. 3. Sketch map of Northeast Cape Fear River showing outcropping formations (from N. C. Geol. and Econ. Surv., Vol. 3).

spatula Forbes was found to underlie the Pliocene Waccamaw formation (fig. 20).

Black River (Sampson County)

On Black River, typical exposures of Peedee materials have been observed between Goff Landing (41½ miles above Wilmington) and Point Caswell (36 miles above Wilmington). According to Stephenson, these are low bluffs never more than 2 or 3 feet above water level. A few fossils were reported from Goff Landing and Sparkleberry Landing (fig. 2).

Northeast Cape Fear River (Pender and New Hanover Counties)

The banks of the Northeast Cape Fear River exhibit the Peedee formation from Deep Bottom Bridge, 75 miles above Wilmington at numerous places to a point about 44 miles above that city and again at Hilton Park near Wilmington.

The farthest upstream exposure of Peedee is at Deep Bottom Bridge on the left bank of the river, ½ mile northwest of Sloan, Duplin County. A nearly vertical 37-foot bluff exposed the following section (October 15, 1945):

	Feet
Red and white sand with thin clay streaks; indurated layer 6 feet above base	15
Black clay with fine interbeds of sand	2
Red and white fine sand; thin clay layers	7
Dark clayey glauconitic sand	1
Dark drab plastic clay	5
Coarse sand	½
Dark green argillaceous glauconitic sand; conspicuous bed of *Ostrea subspatula* Forbes; foraminifera	6½
River level	

About 1 mile upstream, at Chinquipin Bridge, Eocene (Castle Hayne) fossils are exposed. While the contact was not observed, it probably occurs between these two exposures. The entire section at Deep Bottom Bridge is referred to the Peedee with the upper 30 feet being non-marine, or at least non-fossiliferous.

Stephenson lists the following fossil localities along the Northeast Cape Fear River:

Below Deep Bottom Bridge, left bank
67½ miles above Wilmington, right bank
66½ miles above Wilmington
Johnson's Cove (61¾ miles), right bank
Milepost 61, left bank
58¾ miles above Wilmington, right bank
Jackson's Hole, near milepost 58—This is near South Washington (= Watha) where Hodge and Lyell

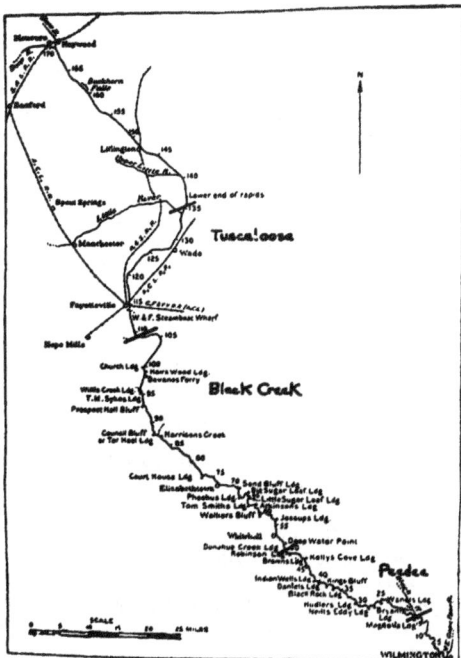

Fig. 4. Sketch map of the Cape Fear River showing outcropping formations (from N. C. Geol. and Econ. Surv., Vol. 3).

obtained many Cretaceous fossils about 1840. Excavations in the vicinity of Watha today reveal only Miocene and Eocene fossils

Crooms Bridge, milepost 56
44 miles above Wilmington (Peedee greensand, no fossils)

Between this point and Hilton Park, Wilmington, no exposures of Peedee rise above river level. At Lanes Ferry and Castle Hayne, Cretaceous fossils are mixed with the Eocene in numerous quarries. Apparently, the Peedee beds are not far below the surface; it is possible that there was a mechanical mixing of the two faunas perhaps during the Eocene. Typical Cretaceous fossils such as *Pholadomya littlei* Gabb and *Cassidulus* sp. were found at Castle Hayne.

The lowest point on the Northeast Cape Fear River where Peedee fossils have been found is at Hilton Park near the northern edge of the city of Wilmington where a section was reported by Stephenson (Vol. 3: 155–156).

Neuse River (Lenoir County)

Peedee beds occur interruptedly along the Neuse River from near Williams Landing, 52 miles above New Bern (about 2 miles above Kinston) to the mouth of Contentnea Creek. Fossils were reported by Stephenson at the following places:

Near Williams Landing, 2 miles above Kinston
Near Kinston
12 miles below Kinston, left bank

This latter locality was visited on May 13, 1942. The Peedee greensand was overlain by Castle Hayne shaly clay. The locality was on the farm of J. H. Cannaday, near the site of an old ferry across the Neuse River.

In 1945 an attempt was made to find the Eocene-Peedee contact reported by Stephenson (1912: 157) near Kinston (34⅔ miles above New Bern) at the site of an old ferry. However, the upper part of the section was no longer exposed. Shell fragments and foraminifera were obtained from the Peedee at this point.

A nearby locality, left bank of Neuse River, 2 miles east of Kinston, showed glauconitic marly sand, but no fossils.

Contentnea Creek (Greene and Pitt Counties)

The bluffs of Contentnea Creek expose Peedee beds at several places between a point about 6½ miles below Snow Hill, Greene County, to Grifton, Pitt County. The bluffs are all low. The best fossil locality was a point near milepost 20, near Hookerton. A few fossils were also found at milepost 6, 1½ miles above Grifton.

Tar River (Pitt County)

Near Greenville.—The cephalopod *Belemnitella americana* Morton was listed by Conrad in 1871 from excavations near Greenville. From this it is deduced

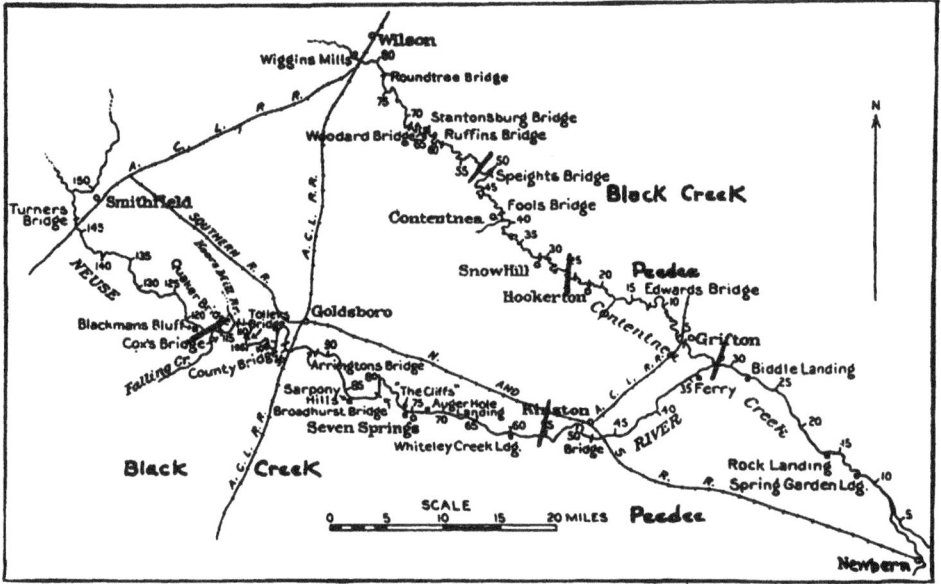

Fig. 5. Sketch map of Neuse River and Contentnea Creek showing outcropping formations (from N. C. Geol. and Econ. Surv., Vol. 3).

that the Peedee beds occur at no great depths in this region.

Columbus County

Lake Waccamaw.—Peedee fossils have been found on the beach along the north shore of Lake Waccamaw. The bluffs here consist of Miocene (Duplin) and Pliocene (Waccamaw) deposits, but it is thought that the Peedee beds occur only a short distance beneath the water level. Some of the Peedee fossils have been reworked into the basal conglomerate of the Miocene.

Freeland.—Greenish stone with shell fragments and shark teeth occurs along the Waccamaw River on the farm of W. A. Mintz near the Columbus-Brunswick

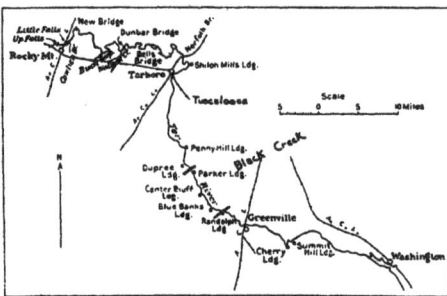

Fig. 6. Sketch map of Tar River showing outcropping formations (from N. C. Geol. and Econ. Surv., Vol. 3).

county line in the "Freeland section." Crocodile bones were obtained from this region by Asa R. Inman in 1937 and presented to the North Carolina State Museum.

Bolton.—A well at the Waccamaw Land and Lumber Company reported by Stephenson (Vol. 3: 160) shows the Peedee extending from a depth of 50 feet to at least 220 feet.

Pender County

St. Helena.—A well for the Carolina Trucking Company, 2½ miles south of Burgaw showed the Peedee between the depths of 60 and 220 feet (Stephenson, Vol. 3).

New Hanover County

Stephenson (Vol. 3: 162–170) recorded several wells in the vicinity of Wilmington. The thickness of the Peedee was given as follows:

Castle Hayne— 44 to   365 feet from the surface
Hilton Park—   0 to   720 feet from the surface
Fort Caswell— 254 to 1,140 feet from the surface

CORRELATION

The fauna of the Peedee formation is equivalent to the *Exogyra costata* zone of Stephenson which comprises the Monmouth group in Maryland and New Jersey. In New Jersey the Monmouth group has been subdivided into the Red Bank, Tinton, Navesink, and Mount Laurel formations.

To the south the Peedee also occurs in South Carolina and is correlated in part with the Ripley formation of the

Alabama and Mississippi and with the Navarro group of Texas.

Stephenson subdivided the *E. costata* zone and recognized a basal zone known as the *E. cancellata*. He has traced this zone all the way from Atlantic Highlands, New Jersey, to the State of San Luis Potosi in Mexico (Stephenson, 1933). In North Carolina this zone marks the lower part of the Peedee formation and has been recognized from the following localities:

Pits on land of J. F. Brooks, 2 miles east of Grifton, Pitt County

Near Sparkleberry Landing, Black River, Bladen County

On Cape Fear River at Black Rock, Daniels, Indian Wells, Kelly's Cove, Robinsons, and Donahue Creek Landing, Bladen County

North Shore of Lake Waccamaw, Columbus County

The *E. cancellata* zone is deeply buried in South Carolina but is present along most of the Gulf Coastal Plain. In New Jersey, the zone is included in the Mount Laurel formation.

TABLE 4

CORRELATION OF PEEDEE FORMATION WITH CONTEMPORANEOUS FORMATIONS IN NEW JERSEY

NORTH CAROLINA			NEW JERSEY
PEEDEE	*Exogyra costata* zone		Tinton Red Bank
			Navesink
		E. cancellata zone	Mt. Laurel

FOSSILS FROM PEEDEE FORMATION

The fossils from the Peedee formation have been described and figured by Stephenson (1923). A few of the more characteristic species are shown in figure 61 of the present report. Unfortunately, most of the localities recorded by Stephenson and other earlier workers are no longer accessible for collecting, having been obliterated by slumping, floods, or other changes. It is therefore difficult to obtain a good collection of Peedee fossils. A few specimens may be found along the Cape Fear River and a few others in excavations at Wilmington and Castle Hayne. Better collecting from this formation can be obtained from the banks of the Intra-Coastal Canal at Myrtle Beach, Horry County, South Carolina.

Probably the most conspicuous species from this formation are: *Exogyra cancellata* Stephenson, *E. costata* Say, *Anomia argentaria* Morton, *Ostrea subspatula* Forbes, *Cardium spillmani* Conrad, *Pholadomya littlei* Gabb, *Gryphaea vesicularis* Lamarck.

EOCENE

LOWER AND MIDDLE EOCENE

No deposits of Lower or Middle Eocene were reported from North Carolina in the 1912 report (Vol. 3), although the presence of some Eocene outcrops near Garner was indicated on the accompanying map. Recently, evidence has accumulated for the presence in North Carolina of Eocene deposits older than the Castle Hayne formation. While it has been impossible to correlate these isolated localities with definite formations, it is thought that they may represent Claiborne or Wilcox time, or both. They may be equivalent in part to the Black Mingo formation of South Carolina or the Pamunkey group (Aquia and Nemjemoy formations) of Virginia. Cooke (1936: 155) reported the presence of Black Mingo fossils at Lillington, Harnett County, North Carolina.

EXTENT

In North Carolina Lower or Middle Eocene deposits are known only from discontinuous patches near the Fall Line in Harnett, Hoke, Wake, and Johnston Counties. Middle Eocene fossils have also been recognized in wells at Williamston, Martin County, Cape Hatteras, Dare County, and elsewhere in northeastern North Carolina.

UNDERLYING FORMATIONS

These deposits rest unconformably on the Tuscaloosa formation (Harnett and Hoke Counties) or on the bed rock (Johnston and Wake Counties) and on Upper Cretaceous material in the wells in northeastern North Carolina.

OVERLYING FORMATIONS

Near Lillington, Harnett County, these deposits are overlain unconformably by the "High Level gravels" of Pliocene age. Otherwise, at the few outcrops they are at the surface. In the Williamston well, the Castle Hayne (Late Eocene) is missing; in the Hatteras well and elsewhere in eastern North Carolina, the Middle Eocene lies beneath the Castle Hayne formation.

LITHOLOGY

When exposed at the surface, the deposits of this age usually consist of consolidated rock in the form of ironstone, sandstone, quartzite, or a peculiar type of limestone locally known as "Fuller's Earth," but not the true rock of this name. Under the surface in Eastern North Carolina, as demonstrated by a study of well samples, Middle Eocene deposits consist of sands, clays, and marls.

THICKNESS

In North Carolina Lower Eocene rocks are known only from very thin isolated patches and are never more than a few feet in thickness.

Harnett County

1. *Lillington.*—A deposit of iron sandstone at the old pits of the Cape Fear Gravel Company, 2 miles northwest of Lillington, was tentatively referred to the Black Mingo formation by Cooke. A few fossil casts were observed but the only identifiable species was *Ostrea arosis* Aldrich (fig. 19).

The Eocene is overlain by the "High Level gravels," but the exact contact between the weathered Eocene and the Pliocene is difficult to determine.

2. *Spout Springs.*—On the Sprunt or Highland Farm, 4 miles northeast of Spout Springs and 1½ miles south of Barbecue Church, there are deposits of a chalky limestone. The limestone is just below the surface at the top of a low hill. Fossils are numerous although few species are represented. They apparently do not correlate with any other Eocene formation in North Carolina and the locality is tentatively referred to the Black Mingo formation. *Pinna harnetti* Richards is the most common fossil.

A nearby hill on the Sprunt farm is capped with iron sandstone, similar to that exposed at Lillington; no fossils were observed.

Hoke County

3. *McCain.*—At the top of several low hills near the Sanatorium there can be seen thin layers of iron sandstone similar to that from Lillington and Spout Springs. These overlie clay of the Tuscaloosa formation.

Kerr (1875: 150) reported fossil shells and echinoderms in calcareous sandstone on top of a hill in the southeastern part of Moore County. The county boundaries have been changed in recent years, and it is highly probable that Kerr's locality is now in Hoke County.

Johnston County

4. *Clayton.*—Outliers, questionably referred to the Claiborne Eocene, occur along Highway 70 just beyond the Wake-Johnston County line and 3 miles west of Clayton. About 200 yards north of the road there can be seen numerous slabs of sandstone and quartzite containing casts of gastropods and pelecypods. These slabs rest unconformably on Carboniferous (?) granite. In addition, there are slabs of iron conglomerate similar to those from Lillington. The fauna of the Clayton locality was recently recorded and figured (Richards, 1948). Because of the lack of typical Castle Hayne species and the presence of several species of Middle Eocene affinity (*Spirulaea rotula* Morton, *Venericardia planicosta* Lamarck var. and *Meretrix ovata* Rogers), it is believed that the locality is of Middle Eocene age, probably correlated with the Claiborne deposits of the Gulf Coast (fig. 18).

Wake County

5. *Auburn.*—The U. S. National Museum has a few fossils obtained by C. W. Cooke and L. W. Stephenson from a road ditch near the railroad 1.6 miles west by north of Auburn on the road to Garner (U.S.G.S. 1/388).

Kerr (1875: 150) also records Eocene "shell conglomerate" along the railroad track 7 miles east of Raleigh.

6. *Meredith College.*—According to Dr. John M. Parker (personal communication) there is an outlier of sediment questionably referred to the Eocene in gulleys in a stream valley half way between Meredith College and the Dairy Barns to the west. This lies on the bedrock (Pre-Cambrian?).

7. *Crabtree Park.*—Quartzite similar to that exposed near Clayton (locality 4) occurs along the banks of Crabtree Creek in the Recreation Area, 11 miles west of Raleigh. The only identifiable fossil was *Amaraulina* sp.

Martin County

8. *Williamston.*—Greensand containing shells of *Gryphaeostrea vomer* were found between 100 and 110 feet in a well. This species is characteristic of the Pamunkey Group (Wilcox Eocene) of Virginia.

Elsewhere in eastern North Carolina, Eocene microfossils presumably older than the Castle Hayne have been found in wells usually beneath definite Castle Hayne species. Since these have not yet been fully studied, it is impossible to make any definite correlations.

Halifax County

9. *Thelma.*—A road cut on highway 258, about 2 miles east of Thelma yielded some poorly preserved fossils tentatively referred to the Eocene.

CORRELATION

Because of the small number of species and the discontinuous distribution of the Middle Eocene deposits in North Carolina, it is difficult to correlate the deposits with those outside the State, or to say that the localities here referred to the Middle Eocene in North Carolina are of the same age. It does, however, appear that there are deposits in North Carolina of Eocene age older than the Castle Hayne. The fact that they occur so far inland does not favor a correlation with the Castle Hayne. Furthermore, the presence of *Ostrea arrosis* at Lillington, *Venericardia planicosta, Spirulaea rotula,* and *Meretrix ovata* at Clayton, *Gryphaeostrea vomer* at Williamston, and the microfossils in other wells favors a correlation with the Claiborne or Wilcox of the Gulf Coast.

FOSSILS FROM THE LOWER AND MIDDLE EOCENE

See Richards (1947a, 1948) for discussion and illustrations of the fossils referred to the Lower or Middle Eocene of North Carolina. Some of these species are shown in figures 62, 63. The fossils are usually poorly preserved. The following are typical (numbers refer to localities described above):

Gastropoda

Amaurellina sp. 7
Acteocina sp. 4
Natica sp. 4
Turritella spp. 4

Pelecypoda

Grypheostrea vomer Morton 9
Meretrix ovata Rodgers 4
Pecten sp. 4
Pecten membranosus Morton 2
Pinna harnetti Richards 2
Psammobia cf. *eborea* Conrad 4
Ostrea sellaeformis Conrad 4
O. cf. *thirsae* Gabb 4

Echinodermata

Schizaster armiger Clark 2

Annelida

Spirulaea rotula Morton 4

CASTLE HAYNE FORMATION

NAME

Named by Miller (Vol. 3) in 1912 from the village of Castle Hayne in New Hanover County, North Carolina. It was regarded as the youngest Eocene deposit in the State and overlying the Trent formation. At that time no extensive study had been made on the fauna. In 1926 Kellum made a careful study of the fauna and showed that the Castle Hayne correlated with the Santee limestone of South Carolina, the Ocala of Florida, and the Jackson formation of Mississippi. Furthermore, it was shown that the Trent marl actually is younger than the Castle Hayne and apparently overlies it. At the same time, it was shown that numerous localities referred by Miller to the Trent, for example along the Trent river, actually contained Castle Hayne fossils. The mapping in the present study has closely followed that of Kellum.

EXTENT

The Castle Hayne extends from Wilmington northeast to the Neuse River and just beyond. Outcrops are known from New Hanover, Pender, Duplin, Onslow, Jones, Lenoir, Southeast Wayne, southern Pitt, and Northwestern Craven Counties.

UNDERLYING FORMATIONS

The Castle Hayne formation lies unconformably on the Cretaceous. This contact can be observed at Castle Hayne, New Hanover County, near Chinquipin Bridge, Pender County, and along the Neuse River 10 miles below Kinston.

OVERLYING FORMATIONS

Locally in New Hanover and Pender Counties, the Castle Hayne is overlain by thin deposits of Waccamaw age (as at Wilmington, Watha, etc.). Elsewhere it is overlain by local deposits of Duplin marl or by Pleistocene terrace material. The overlying Duplin deposits and the contact are best exposed at the Natural Well, near Magnolia in Duplin County. In this sink hole, a contact between the Eocene and Miocene can be observed.

LITHOLOGY

In many places the Castle Hayne formation consists of a finely broken calcareous marl, which is often between 90 per cent and 100 per cent $CaCO_3$. Numerous occurrences of this marl were listed by Miller (1912) and Kellum (1925). Locally, the Castle Hayne is much more consolidated and near Wilmington and Castle Hayne has been quarried for crushed and dimension stone. At Pollocksville, a phase of the Castle Hayne occurs which is almost entirely composed of fragile whole shells of the giant oyster *Ostrea georgiana*. Another local phase of the Castle Hayne occurs along the Neuse River about 10 miles below Kinston. Here it occurs as a lightweight unfossiliferous gray clay, somewhat resembling Fuller's Earth.

DETAILED SECTIONS

Pitt County

1. *Three miles east of Quinnerly.*—On the Green farm, 2 miles north of the Neuse River on the Pitt-Craven county line along highway 118, occur several marl pits. The marl is bluish-white and is of high lime content. The farm was formerly owned by T. J. Falkner and as such was recorded by Kellum (U.S.G.S. 10627).

Perfect fossils are rare, but the following have been identified: *Cassidulus carolinensis craveni* Kellum; *Ostraea sellaeformis* Conrad; *Pecten membranosus* Morton; *P. cushmani* Kellum; *P. cookei* Kellum; *Anomia* sp.

2. *Clayroot Swamp.*—Marl, similar to the above, was dredged from drainage canals just south of Quinnerly Bridge, in Clayroot Swamp on the road from Fort Barnwell to Shelmerdine. The fauna was essentially the same as the above (U.S.G.S. 8169).

Craven County

3. *Maple Cypress.*—Shallow pits were dug for marl about ½ mile south of the general store. The material at the side of the pits was chiefly Castle Hayne marl with typical *Pecten cookei*. In addition, there was some rock carrying poorly preserved casts of mollusks which resembled those of the Trent formation. There were also a few fragments of whale bones lying on the surface which are probably of Yorktown age. Typical Yorktown fossils have been reported from this locality, but none were observed from the pits examined on the

present survey. It is, of course, impossible to give a section at this place; however, the relationship is something as follows:

Thin layer of sand with whale bones........Yorktown
Indurated rock.............................Trent (?)
White shell marl, high in lime.............Castle Hayne

4. *Biddle Landing.*—Some marl was formerly dug on the right bank of Neuse River due north of the town of Fort Barnwell. The deposit is about 20 feet thick and consists of a cemented marl at the river's edge overlain by some limey sand with numerous shell fragments, also of Castle Hayne age.

5. *Fort Barnwell.*—Marl pits on the property of Z. B. Broadway, 1 mile north of Fort Barnwell revealed some consolidated rock probably of Castle Hayne age beneath a deposit of Yorktown shell marl. For further details, see under Yorktown (p. 26).

6. *Cannon's Marl Pit.*—This marl pit is situated near the bridge over the Neuse River, a mile northeast of Fort Barnwell. Here also, the Castle Hayne is overlain by loose Yorktown shell marl. Characteristic Jackson (= Castle Hayne) species were identified. Similar Eocene rock outcrops occur at very shallow depths at various places in the Fort Barnwell area.

7. *Turkey Trap Farm.*—Castle Hayne marl was dug in 1942 for agricultural and road building purposes from this farm, which is located on the Dover Road, just east of the Lenoir County line, four miles northeast of Dover on Halfmoon Creek. The fossils were numerous and fairly well preserved.

Lenoir County

8. *Whiteley Creek.*—Marl has been reported from the Dave Wilkins and Outlaw Plantations, near the junction of Whiteley Creek and the Neuse River. According to Berry and Cushman (1921: 107) the marl contained 74.05 per cent $CaCO_3$ and 0.5 per cent phosphoric acid.

9. *Four miles south of Kinston.*—Berry and Cushman report old marl pits along Mill Branch containing 70–88 per cent $CaCO_3$. Bryozoa were said to be present.

10. *Cannaday Farm.*—On the farm of J. H. Cannaday, 12 miles below Kinston on the left bank of the Neuse River near the location of an old ferry, shallow pits revealed the presence of a light shaly clay, somewhat resembling "Fuller's Earth." This lay on top of greensand marl of the Peedee formation.

Wayne County

11. *Broadhurst Bridge.*—There is an outcrop of bluish-white shell marl, about 3 feet thick where a small stream empties into the Neuse River at the bridge of Route 111 between milepost 82 and 83. *Ostrea trigonalis* and *Pecten cookei* are most common (U.S.G.S. 10625).

12. *Mount Olive.*—Berry and Cushman (1921) listed marl pits containing marl of Castle Hayne age on the

Flower Farm and elsewhere in the vicinity of Mount Olive.

Jones County

13. *Two and One-half Miles West of Dover.*—Former pits were noted by Kellum on the J. L. Bryan farm. *Ostrea sellaeformis, Pecten membranosus* and *P. cookei* were collected from the banks (U.S.G.S. 10632).

14. *Beaver Creek.*—Two miles northwest of Wilcox Bridge over the Trent River, Miller (Vol. 3: 184) collected some bryozoa and other fossils (U.S.G.S. 7803).

15. *One-half Mile Northwest of Wimsatt.*—Marl has been quarried and used locally on the farm of D. W. Dudley on the right bank of Little Chinquipin Creek (U.S.G.S. 10631).

16. *Comfort.*—The following section was recorded by Kellum (1925: 10) on the right bank of the Trent River, on the farm of Miss Sally Simmons, ¾ mile southwest of Comfort depot (U.S.G.S. 10630).

	Feet
Pleistocene: Soft, unconsolidated gray sand	3
Eocene (Castle Hayne marl):	
Bluish-gray sticky clay, unfossiliferous	3
Yellow ground-shell marl, with sandy matrix containing Bryozoa and *Pecten deshayesii* Lea in abundance and *Periarchus lyelli* (Conrad) rarely	5
Hard yellow impure limestone, with numerous casts of fossils	5
Light-blue soft marl that turns yellow on exposure to the air; contains a few poorly preserved shells	9

17. *Five Miles Northwest of Trenton.*—Marl was recently dug on the property of Carl Thigpen at this point. Similar marl outcropped along the Trent River in this vicinity. Perfect fossils were scarce, although some bryozoa was obtained from the pit.

18. *Trenton.*—Large shells of *Ostrea georgiana* are in the North Carolina State Museum obtained from ½ mile east of Trenton.

19. *One mile East of Pollocksville.*—Large specimens of *O. georgiana* are common in a pit used by the highway department.

20. *Pollocksville.*—There is an extensive marl pit at the Atlantic Coast Line Railroad station. *O. georgiana* is abundant, and occurs in a solid sandy bed about 15 feet thick. It is immediately across Trent River from the locality mentioned by Miller (U.S.G.S. 7809) (fig. 24).

21. *Maysville.*—Excavations of the Raleigh Granite Company in 1942 passed through the Trent marl to the underlying Castle Hayne. While the majority of the fossils from the dumps have been correlated with the Trent formation a few species which probably came from greater depths, may be from the Castle Hayne formation.

Onslow County

22. *Belgrade.*—On the opposite side of the Whiteoak River from the Maysville locality mentioned above (21) are the extensive quarries of the Superior Stone Company. The limestone has been quarried for road metal

and concrete aggregate and was used extensively by the Camp Lejeune Marine Base and other nearby military installations during World War II. Here, as at Maysville, the majority of the fossils came from the Trent formation, although specimens from deeper parts of the excavation indicate a Castle Hayne age. The presence of the two formations was confirmed by a study of microfossils from auger borings.

23. *Two Miles North of Jacksonville.*—Kellum reports marl dug from the Sabiston Farm, 2 miles north of Jacksonville (U.S.G.S. 10637). No pits could be observed at the present time.

24. *Ten Miles Northwest of Jacksonville.*—Kellum recorded a large outcrop of consolidated yellow, ground shell marl, about 10 feet thick, extending horizontally around the nose of a low hill on the farm of J. M. Thomas (U.S.G.S. 10636).

25. *Richlands.*—Limerock with bryozoa was dredged from the New River at the crossing of Route 258. Kellum recorded several former marl pits in this immediate vicinity (U.S.G.S. 10635).

Duplin County

26. *Natural Well.*—This is a natural sink hole located 2 miles southwest of Magnolia, and is the type locality of the Duplin formation of Miocene age. This shell marl rests on light green sand which has been variously regarded as of Eocene or Miocene age. Huddle (1940) examined some microfossils from the locality and determined that they were of Jackson age, and hence are probably correlated with the Castle Hayne (Eocene). This seems logical inasmuch as this formation occurs at several other places near Magnolia.

27. *Two Miles South of Magnolia.*—Kellum reported bluish white marl, with numerous fossils, from pits on the farm of B. D. Johnson (U.S.G.S. 10622).

28. *Rose Hill.*—Similar marl was dug on the farm of A. L. Bland, 1½ miles northeast of Rose Hill Station (U.S.G.S. 10621).

29. *Two Miles South of Kornegay.*—Marl was being dug in 1942 at the Hamp Williams pit on North Carolina highway 111. The marl is bluish white and rich in lime. Perfect shells are rare, although a few specimens of *Pecten cookei* and *Ostrea trigonalis* were collected.

30. *Three Miles South of Kornegay.*—And about 5 miles west of Pink Hill on highway 111, there was formerly a marl pit on the property of B. F. Smith. The farm is now owned by Mrs. Nell Ruth Smith and has been leased to William Houston (1942). A few abandoned marl pits were seen about ½ mile west of the road.

31. *Cedar Fork Swamp.*—Several marl pits were noted along the side of the road at this small community about 5 miles east of Beulaville. These were rich in lime and had apparently been dug recently (near U.S.G.S. 10624) (fig. 22).

32. *Chinquipin Bridge.*—About 100 yards below this bridge across the Northeast Cape Fear River and on the left bank, there is an outcrop of hard limestone about 4 feet thick. Fossil casts are abundant but indistinct. This grades upward into a less consolidated marl which has been dug for agricultural purposes (fig. 23).

Pender County

33. *Watha.*—Marl has been dug on the farm of A. A. McMillan, 1 mile southeast of Watha where the road crosses Lewis Creek. The majority of the marl was from the Duplin formation, but it apparently overlay the Castle Hayne.

This is near the place formerly known as South Washington, where Cretaceous fossils were collected many years ago. Apparently, older pits went deeper and penetrated through the Miocene and Eocene formations into the underlying Cretaceous (Peedee?).

34. *Maple Hill.*—Marl, rich in lime, has been dug on the farm of O. V. Wooten, 1 mile south of Maple Hill post office (U.S.G.S. 10618).

35. *Moores Creek.*—Two miles west of Angola Post Office and 3 miles northeast of Maple Hill. Marl pits have been recently dug on the farm of Caesar Jones. These are adjacent to the former pits on the property of J. L. Fisher (U.S.G.S. 10619).

36. *Rocky Point.*—Stephenson recorded Castle Hayne fossils from the French Brothers Quarry at old Rocky Point (now Lanes Ferry) on Northeast Cape Fear River, 3 miles east of Rocky Point station. The following section was given:

	Feet
Surficial brown sandy clay	5
Eocene (Castle Hayne marl):	
Dark yellow to greenish disintegrated limestone stained more or less with iron	.5
Soft white limestone or marl with many fossils	2
Hard white limestone	2.5
Conglomerate composed of phosphatic pebbles or nodules up to 2 inches in diameter, shark teeth, and mollusks, cemented with calcium carbonate	2

Cretaceous (Peedee) beds underlie the Eocene and some such fossils have been reworked into the basal beds. At the time of our visit (1941) no identifiable fossils were obtained.

37. *New Rocky Point Quarries.*—Stephenson (1927) listed some Cretaceous fossils (Peedee) that probably reworked and mixed with typical Castle Hayne species at New Rocky Point Quarry, one mile northeast of Rocky Point station. The specimens were obtained from the dumps.

38. *Bob Bourdeau Tract.*—On the left bank of the Northeast Cape Fear River, 3½ miles above Castle Hayne bridge, Kellum reported an outcrop of loosely consolidated yellow earthy marl, 5 to 7 feet thick bearing Castle Hayne fossils.

New Hanover County

39. *Castle Hayne.*—The rock quarry ½ mile south of the railroad station is the locality where the majority of the Castle Hayne fossils have been collected. The following section was noted (after Kellum) :

	Feet
Pleistocene: Gray and yellow loose sand of varying texture, about	6
Eocene (Castle Hayne marl):	
White and grayish-white marl containing some fossils	2–2½
Coarse conglomerate of green and black phosphate pebbles with sandstone and quartz pebbles cemented by lime	3
Cretaceous: Brownish-yellow impure limestone made up almost entirely of casts of marine mollusks	4–6

Excellent fossils were obtained from this quarry in 1942, 1944, and 1946.

There is some question whether the Cretaceous is actually in place or whether some Cretaceous fossils have been reworked and incorporated into the Eocene. Cretaceous fossils such as *Pholadomya littlei* Gabb and *Cassidulus* spp. are common. The most typical Eocene species are: *Crassatellites alta* (Conrad) ; *Meiocardia carolinae* Harris; *Terebratula wilmingtonensis* L. and S.; *T. lachryma* Morton. Sharks' teeth are especially abundant. Similar pits occur in the immediate vicinity (fig. 23).

40. *Wilmington.*—The City Rock Quarry on the east side of Wilmington near Smith Creek was formerly an excellent place to collect Eocene fossils. The quarry is now entirely filled with water and it is impossible to obtain good material. The Eocene is overlain by a shell rock containing *Pecten, Plicatula, Balanus,* etc., but it cannot be determined whether this is Pliocene (Waccamaw) or Miocene (Duplin).

CORRELATION

Miller (Vol. 3) referred the Castle Hayne to the Upper Eocene, later than the Trent marl. However, at that time little work had been done on the respective faunas. Cooke (1916) dated the Castle Hayne a little closer when he said (p. 111) that certain species from the Ocala limestone of Florida "apparently occur also in the Castle Hayne limestone at Wilmington, North Carolina, which is of Jackson age." Canu and Bassler (1920) studied the bryozoa from the Castle Hayne and recognized 214 species and varieties, of which 169 were new and 88 were limited to the Castle Hayne. Through this study the formation was assigned to the middle of the Jackson horizon of the Eocene and correlated with the Cooper marl of South Carolina, the *Ostrea georgiana* zone at the base of the Barnwell formation of South Carolina, and the Trivola tongue of the lower part of the Ocala limestone of Georgia.

Kellum (1926) restudied the mollusks and echinoderms of the Castle Hayne formation and confirmed the

Jackson correlation. Of the 53 species, he noted the following relationship :

Limited to Castle Hayne marl	30
Limited to Jackson horizon in other regions	9
Limited to Claiborne horizon in other regions	1
Known from Jackson and older horizons	11
Claiborne to Tampa	1
Horizon doubtful	1
	53

Huddle (1940) has recently determined some microfossils of Jackson age from the Castle Hayne formation at the Natural Well.

For further lists and illustrations see Kellum (1926).

FOSSILS FROM CASTLE HAYNE FORMATION

For further lists and illustrations see Kellum (1926).

Gastropoda

Cypraea sp. 39
Fusinus abruptus Tuomey 39

Pelecypoda

Crassatella alta Conrad 39
C. wilcoxi Brown & Pilsbry 39
Meiocardia carolina Harris 39
Ostrea trigonalis Conrad 11, 29
O. georgiana Conrad 20
O. spp.
Pecten membranosus Morton 1
P. cookei Kellum 1, 11, 29
P. deshayesii Lea 39
P. spp.
Plicatula filamentosa Conrad 39

Brachiopoda

Terebratula wilmingtonensis Lyell & Sowerby 39
T. lacryma Morton 39

Echinodermata

Cidaris pratti Clark 39
Periarchus lyelli Conrad 15, 39
Cassidulus carolinensis Twitchell 1, 39
Linthia wilmingtonensis Clark 39

Bryozoa

Many species

OLIGOCENE

Until very recently, Oligocene deposits were not reported from the Atlantic Coastal Plain north of the Georgia-South Carolina boundary. However, some samples from a water well at Camp Lejeune, Onslow County, North Carolina, were studied by McLean (1947) and a microfauna tentatively correlated with the Oligocene was identified between the depths of 60 and 95 feet. The presence of an Oligocene deposit in other wells in eastern North Carolina has also been suggested (McLean, 1947; Swain, 1947, and Miss Doris Malkin in personal communication).

MIOCENE
TRENT FORMATION
NAME

The formation was named by Miller from the Trent River in North Carolina along which stream it is exposed. Miller gave no type locality, but mentioned that it occurred along the Trent River from Trenton, Jones County, to near the junction of the Trent and Neuse Rivers. Very few fossils were reported and the formation was erroneously regarded as being of Eocene age lying beneath the Castle Hayne, although the contact had not actually been observed.

Kellum restudied the Trent and Castle Hayne marls (1925) and from a study of numerous fossils established that the Trent was of Lower Miocene age. Of the twenty-three localities listed by Miller as Trent, seven failed to yield any fossils, and seven proved to be Castle Hayne. It was therefore necessary to redefine the boundaries of these formations. The name Trent is retained for the formation, although it extends along the Trent River only from its junction with the Neuse to just below Pollocksville; the deposits along the Upper Trent river are Castle Hayne.

EXTENT

The Trent formation is exposed east of the Castle Hayne in a narrow north and south belt between the New and Neuse Rivers. It is exposed along the New River at Jacksonville, near the White Oak River in the vicinity of Silverdale, along the Trent River from its junction with the Neuse to a point about 3 miles below Pollocksville, and along the Neuse River from New Bern to at least Rock Landing, about 16 miles upstream from New Bern. It is therefore found in Onslow, Jones, and Craven Counties.

UNDERLYING FORMATIONS

The Trent formation is underlain by the Castle Hayne. Although the contact has not been observed above the surface, it is noted in several wells and in the quarries at Belgrade. There is apparently a considerable unconformity.

OVERLYING FORMATIONS

Near Silverdale, the Trent is overlain unconformably by the Croatan sand of Pliocene age. At the Askew marl pits, 2 miles northwest of Silverdale, both formations are exposed but it is impossible to determine a clear contact. There has apparently been some mechanical mixing of the faunas. The Trent marl is reported from wells south of New Bern immediately under the Croatan sand. However, north of New Bern, the Trent is overlain by the Yorktown formation, late Miocene, at least as far upstream as Rock Landing. Questionable Trent stone was found near the Neuse River in marl pits at Maple Cypress, Craven County.

The chief marl at this place was Castle Hayne, which was overlain by a thin deposit of Yorktown shell marl. If the Trent occurs at this locality, it must be a very thin layer lying between the Castle Hayne and the Yorktown.

LITHOLOGY

The Trent is usually a consolidated shell rock which is often used as building stone. Numerous houses in the vicinity of Jasper have foundations of this stone and the walls of the cemetery in New Bern are also of this material. Fossils are numerous, but usually in the form of casts.

Locally the Trent is unconsolidated and resembles younger marls such as the Yorktown. This phase is best exposed near Silverdale and it is from this region that the best fauna was obtained.

Thickness.—Undetermined, probably less than 200 feet. It is difficult to differentiate the Trent from the Castle Hayne in material recovered from wells.

Dip.—Dips gently toward the ocean.

DETAILED SECTIONS
Craven County

1. *Camp Battle.*—3 miles above New Bern on the Neuse River, ¼ mile above the Pumping Station. A limy sandstone is exposed to about 3 feet above normal river level. Pelecypod casts are common. This rock is covered with a layer of barnacles (*Balanus* sp.) which is probably of Yorktown age. This suggests that the rock ledge was on the ocean shore line during Yorktown time.

2. *Spring Garden.*—At this landing on the right (south) bank of the Neuse River, 3½ miles northeast of Jasper, Trent limestone occurs up to some 4 feet above the normal river level. Casts of shells are common, although the shells themselves have been dissolved. According to Miller (Vol. 3: 179) this stone has been used for buhrstones (millstones).

3. *Jasper.*—Outcrops of Trent limestone are common near Jasper. On the B. S. Dawson farm, about 1.5 miles east of Jasper, there is a small quarry where stone has been dug for road construction. Also, stone has been quarried from this and similar nearby places for use in building the wall of the cemetery in New Bern. Similar stone is exposed in a ditch along the road near the Dawson farm. Casts of *Crassatella* sp. are common (fig. 29).

4. *Rock Landing.*—At this old landing on the Neuse River, about 16 miles above New Bern and 3 miles east of Jasper, can be seen good exposures of Trent limestone. Nearby, in an old marl pit, it is overlain with unconsolidated Yorktown marl.

5. *Maple Cypress.*—Pieces of rock, resembling that from Rock Landing and Spring Garden, were found near a marl pit about ½ mile south of the general store at Maple Cypress. The chief source of marl here was the Castle Hayne formation, so if the above-mentioned

rock is actually of Trent age, it lies on top of the Castle Hayne and beneath a thin mantle of Yorktown. See page 26.

6. *Rock Springs.*—On the right bank of the Trent River, 3½ miles above New Bern, typical consolidated Trent limestone has been observed on the farm of James N. Smith. Casts of *Callista neusensis* Harris are abundant (U.S.G.S. 10657).

7. *Havelock.*—Mansfield (1925) recorded the Trent formation in the "oil" well at Lake Ellis, 4 miles west of Havelock, as occurring between the depths of 278 and 455 feet. It was underlain by what is probably the Castle Hayne.

8. *White Rock Landing.*—On the left bank of Trent River 12 or 13 miles above New Bern, a rock similar to that at Rock Springs rises to 15 feet above the river level. Casts of Trent fossils have been recognized in this rock (U.S.G.S. 7796).

9. *Scott Landing.*—A former quarry of the Atlantic Marl and Lime Fertilizer Company on the left bank of the Trent River showed the following section (after Kellum; U.S.G.S. Station 10656):

	Feet
Unconsolidated white or light yellow sand Pleistocene	1½–3
Ground shell marl with little sand Miocene (?)	12–15
Hard yellow impure limestone made up entirely of casts of marine shells Miocene (Trent)	6–10

10. *Simmons Marl Pit.*—3 miles east of Pollocksville and one mile up the Trent River from the previous locality. Shell marl is abundant, but perfect shells are rare (fig. 27).

Some years ago there were a number of other marl pits along the Trent River, but at the present time this is the only one operating. Harris (1919) recorded *Pecten trentensis* Harris and *P. exilatus* Conrad from the right bank of the Trent River, 6 miles below Pollocksville. At that time the Trent formation was thought to be of Eocene age, but Harris then said that these fossils cast some doubt on that correlation. *P. trentensis* has recently been found at Simmons Pits (1946). Harder rock, similar to that characteristic of the Trent formation at Belgrade, was noted in deeper parts of the Simmons pit.

11. *Maysville.*—Excavations by the Raleigh Granite Company in 1942 revealed an extensive deposit of fossiliferous Trent limestone. The material is similar to that at Belgrade, on the opposite side of Whiteoak River, and is described more fully under that locality (no. 15).

12. *Silverdale.*—Marl has been dug from the Gillette farm, ½ mile southwest of Silverdale. It is from this locality that Kellum obtained the fossils from which he determined the Miocene age of the Trent formation. At these pits, on the south side of Webbs Creek, there is an exposure of yellow rotten shell marl at least 6 feet deep. Kellum described a fauna of 21 species from this locality, of which 16 species were known only from

the Trent formation. Earlier workers had regarded the Trent as of Eocene age, but the fauna described by Kellum suggested a basal Miocene age, closely related to the Tampa limestone of Florida. A certain relationship was also shown to the Vicksburg formation of Oligocene age, but the Miocene relationship seemed closer.

With one exception (*Lyria carolinensis* Kellum), all of Kellum's species were found on the present search; in addition, numerous other species were found bringing the total for the formation to at least 54 species. Eleven of these were described as new in a recent report (Richards, 1943a).

In general these additional species recently collected further confirm the Lower Miocene age of the Trent formation.

One of the most interesting species from this locality is *Donax idoneus* Conrad originally described from the "Coast of North Carolina, probably from a Miocene deposit under the sea" (Conrad, 1872). It is much larger than any other species of *Donax* from any formation along the East Coast of North America, but it is very close to *D. punctostriatus* Hanley from the Gulf of California, the West Coast of Mexico and Central America as far south as Peru. A similar distribution is noted in the case of *Pecten humphreysii* Conrad, an East Coast Miocene species which has a very close relative in the present Pacific Ocean. Furthermore, Lohman has pointed out that the diatom *Aulacodiscus rogersii* (Bailey) Schmidt diagnostic of the Calvert (Middle Miocene) of the East Coast, has been obtained from dredgings off Lower California where it is presumed to be Recent. These examples of a correlation between the East Coast Miocene and the West Coast Recent are highly interesting and add further evidence to the theory of a Tertiary sea across Mexico, probably at the Isthmus of Tehuantepec.

13. *Askew Pits, Silverdale.*—There are several small marl pits on the farm of A. W. Askew, 2 miles northwest of Silverdale and ½ mile east of the highway. The lowest marl beds exposed in a small stream belong to the Trent formation, while the highest beds, 15 feet above the stream level, are part of the Croatan (Pliocene) formation. At most parts of the pits, however, the marl from the two formations has been mixed; probably some of this mixing occurred during the Pliocene. It is therefore difficult to separate the two faunas.

14. *Jacksonville.*—At the foot of New Bridge Street there are some pieces of hard rock with casts of *Cardium, Venus* and *Callista* (= U.S.G.S. 10654).

15. *Belgrade.*—Extensive excavations by the Superior Stone Company on the south bank of the Whiteoak River have revealed a massive deposit of Trent limestone. This recemented shell rock has been found to be peculiarly adapted for stone aggregate, and thousands of tons of this material were used extensively in the construction of the various military establishments

in the vicinity during World War II. The fossils are remarkable for the perfect preservation of the internal structures such as muscle scars, etc. The majority of the species indicate a Miocene age and are referred to the Trent. A few shells and some microfossils from the deeper parts of the excavation suggest an Eocene age and probably indicate the upper part of the Castle Hayne formation. A few of the fossils were described in a recent paper (Richards, 1948).

The quarry is in operation at the present time (1949) (fig. 25).

CORRELATION

To repeat, the Trent formation was described by Miller (Vol. 3) as of Eocene age beneath the Castle Hayne marl. The few fossils then determined were four echinoderms which according to Twitchell (Clark and Twitchell, 1915) seemed "to indicate the Upper Claibornian age of the strata."

Because of the discovery of a well-preserved fauna near Silverdale, Kellum was able to point out the error of this correlation. In a preliminary note (1925) and a detailed study (1926) he showed that the Trent marl actually overlies the Castle Hayne and that its fauna shows a close relationship to that of the Tampa limestone of Florida. He further pointed out that three of the four echinoderms cited by Twitchell have not been recognized elsewhere and therefore are of little correlation value. The fourth (*Echinocyamus parvus* Emmons) has been found in Onslow County associated with Castle Hayne fossils, so probably the original locality was Castle Hayne and not Trent.

Harris (1919) had apparently been dissatisfied with the Eocene dating of the Trent formation, for when he described *Pecten trentensis* from the Trent marl, he remarked (pp. 15–16) it "is so different from anything we have heretofore found in the Eocene of this part of the United States, owing to a lack of well-known species from the same locality, its horizon must at present be considered as doubtful."

Harris also listed *P. elixatus* Conrad from the same locality, a species of Vicksburg (Oligocene) age. This also led Harris to question the Eocene age of the Trent.

Kellum believed that the closest relationship of the Trent fauna was with the Tampa limestone of Florida, of basal Miocene age. However, he pointed out that there was also a relationship with the Vicksburg Oligocene and that therefore the correlation with the Tampa must be regarded as tentative.

The material collected by the present survey, from Silverdale has added 32 species, including 11 new species to the list from the Trent marl. The additions to the fauna confirm the lower Miocene age of the formation.

FOSSILS FROM TRENT FORMATION

For full descriptions and illustrations of the fossils of the Trent formation see Kellum (1926), Richards (1943a) and Richards (1948). Some of the more typical species are listed below.

Gastropoda

Busycon spiniger onslowensis Kellum 12
B. filosum Conrad 12
Calyptraea aperta Solander 12, 15
Crucibulum constrictum Conrad 12
Conus postalveatus Kellum 12
Crepidula fornicata Linne 12
Ecphora quadricostata Say 12
Eupleura harbisoni Richards 12
Murex davisi Richards 12'
M. kellumi Richards 12
Oliva posti Dall 12
Olivella gardnerae Richards 12
Polinices duplicata Say 12
P. hemicrypta Gabb 12'
Potamides silverdalensis Richards 12
Rapana vaughani Mansfield 12
R. gilletti Richards 12
Sinum imperforatum Dall 12
Scaphella stromboidella Kellum 12
Turritella fuerta Kellum 12

Pelecypoda

Arca silverdalensis Kellum 12'
Astarte onslowensis Kellum 12
Anomia onslowensis Richards 12, 15
Antigona lamellacea Kellum 12
Crassatellites mississippiensis silverdalensis Kellum 12
Cardium midyetti Richards 12
C. belgradensis Richards 15
Donax idoneus Conrad 12
Diplodonta alta Dall 12
Glycymeris anteparilis Kellum 12
Macrocallista minuscula Kellum 12
M. tia Kellum 12
Modiolus stuckeyi Richards 15
Ostrea georgiana Conrad 12, 15
Pecten trentensis Harris
Plicatula densata Conrad 12
Phacoides silicatus Mansfield 12
Panope intermedia Richards 15
Venus gardnerae Kellum 12, 15
V. errecta Kellum 12
Venericardia nodifera Kellum 12

CALVERT FORMATION

NAME

Named from the Calvert Cliffs along Chesapeake Bay in Maryland.

EXTENT

This formation has not hitherto been reported from North Carolina. Calvert deposits are known near Richmond, Virginia, and it has been thought that the shore line turned sharply east near that point. Thus any Calvert material in North Carolina might occur in the northeastern or extreme eastern part of the State and probably would be covered by later material. The recent well at Cape Hatteras revealed a thickness of foraminiferal ooze beneath the Yorktown which could possibly be of Calvert age. This zone occurred between 760 and 1,000 feet, and has tentatively been correlated with the Middle Miocene, possibly the Calvert.[3]

Microfossils suggesting a Calvert age have been determined from other wells in eastern North Carolina.

ST. MARY'S FORMATION

NAME

This formation was named by Shattuck (1902) from the St. Mary's River in Maryland.

EXTENT

Miller (1912) mapped this formation as extending over a considerable area in northeastern North Carolina as well as in adjacent Virginia (1912a). Olsson (1917) pointed out that this fauna was sufficiently distinct from the St. Mary's of Maryland to warrant giving it a new name. He therefore proposed the "Murfreesboro" stage with type locality at Murfreesboro, on the Meherrin River, Hertford County, N. C. This stage he regarded as intermediate between the St. Mary's and the Yorktown.

Mansfield (1929, 1936) pointed out that the term "Murfreesboro" was preoccupied and after a restudy of the fauna designated it as Zone 1 or *Pecten clintonius* (lower) zone of the Yorktown.

The true St. Mary's is not known from any outcrops in North Carolina, but according to Mansfield (1926: 171) it "apparently underlies the Yorktown and is revealed only at very low water stages of the streams." However, no confirmation of the presence of the true St. Mary's has been found in the present survey.

In the Cape Hatteras well, the St. Mary's, if present at all, is indistinguishable from the Yorktown.

YORKTOWN FORMATION

NAME

The Yorktown formation is named from Yorktown, Virginia. The name was originally used by Dana (1863) for all Miocene and Pliocene deposits of the Atlantic Coastal Plain, and it was not until 1906 that the formation was definitely described by Clark and Miller from Virginia. It was extended into North Carolina by Miller in 1912, although in that report (Vol. 3) the older phase (Zone 1) was regarded as belonging to the St. Mary's formation. Mansfield (1936a) recognized two zones of the Yorktown in North Carolina, the lower one (Zone 1) or *Pecten clintonius* Zone, approximately equivalent to the "St. Mary's" of Miller and the "Murfreesboro" of Olsson. Later work by MacNeil (1938) has shown still further

[3] Data from Standard Oil Company of New Jersey.

subdivisions of the Yorktown, particularly of Zone 2 (see page 27).

ZONE 1. "MURFREESBORO STAGE"

Although invalid because of previous use of the name for a formation in Tennessee, the term "Murfreesboro" stage is still frequently used for this phase of the Yorktown. It was proposed by Olsson (1917) for the phase of the Miocene which Miller had correlated with the St. Mary's. As stated above, Mansfield has placed this "zone" in Zone 1 or *Pecten clintonius* zone of the Yorktown. *P. clintonius*, however, is not a good index fossil for this bed in North Carolina because it is common only in the extreme northern part of the State.

EXTENT

Zone 1 outcrops in northeastern North Carolina along the Meherrin, Roanoke, and Tar Rivers and sparingly along the Neuse River and elsewhere. It occurs at shallow depths in a broad band from the Virginia-North Carolina line south to the Neuse River. Its western limits are about on a line from Weldon through Halifax, Enfield, Whitakers, Rocky Mount, and Wilson to Goldsboro, although outliers are occasionally found some miles to the west. To the east it dips beneath the later Yorktown (Zone 2) along a line approximately through Winton, Williamston, Washington and Kinston.

UNDERLYING FORMATION

Along the Meherrin River the Yorktown extends below the water level so the underlying formation cannot be determined. Along a belt extending from Northampton and Halifax Counties on the north to Wayne County on the south it is in direct contact with the crystalline rocks of the Piedmont Plateau. This may be seen at Halifax, Rocky Mount, and Wilson. Along the Roanoke River, the Yorktown rests on Tuscaloosa (Cretaceous) while on the Tar River it rests on the Tuscaloosa between Dunbar Bridge and Parker Landing and on the Black Creek (Cretaceous) below that point to the vicinity of Greenville. In the vicinity of Contentnea Creek (as at Maury) it overlies the Snow Hill phase of the Black Creek formation and locally near the Neuse River in Wayne County it is reported overlying the Castle Hayne. In all observed sections there is a considerable unconformity beneath the Yorktown, although in the extreme northern part of the State it may be underlain with only slight unconformity by the St. Mary's. In the Hatteras well the Yorktown, possibly including the St. Mary's, overlies the Calvert (?).

OVERLYING FORMATIONS

Along the Chowan River near Mount Pleasant Landing Zone 1 is overlain by Zone 2 of the Yorktown. Elsewhere it is usually covered by a thin mantle of Pleistocene sand.

LITHOLOGY

Zone 1 consists mainly of dark bluish-green, medium fine, argillaceous sand with many shells. Upon oxidation, it often becomes yellow. Occasionally it is indurated. Marl beds are numerous, and formerly were extensively worked. Sandy phases are common in which case the sand is usually fine, although a pebbly band occurs at the base at some places. Black phosphatic pebbles, derived from the Eocene, are occasionally found. Clay lenses have been observed and a small deposit of diatomaceous earth was reported near Wrendale.

THICKNESS

This is extremely variable and greatest along the Meherrin and Chowan Rivers. In a well at Edenton, North Carolina, Miller stated that the "St. Mary's" (Yorktown 1) was at least 150 feet thick; at Wilson he recorded it as 45 feet thick. The determination of the exact thickness of the various zones of the Yorktown must await careful study of fossils from numerous wells in the region.

DIP

There is a slight dip toward the ocean, less than 10 feet per mile, about equivalent to the present land slope.

ZONE 2

This is equivalent to the entire Yorktown as mapped by Miller in 1912. As stated above the St. Mary's of Miller and the "Murfreesboro" of Olsson are regarded as approximately equivalent to Zone 1 of the Yorktown. For a further discussion of this correlation see page 27.

EXTENT

The upper Yorktown (Zone 2 as a whole) occurs in discontinuous patches across northeastern North Carolina. It is especially well developed along the Chowan, the lower Roanoke, the lower Tar, and the lower Neuse Rivers and their tributaries. It always occurs down stream from Zone 1. Yorktown deposits are not recorded south of the Neuse River. The Miocene of the southern part of the State (Duplin formation) is probably equivalent to the youngest Yorktown. According to Mansfield the fauna from Tar Ferry on Wiccacon Creek in Hertford County, is approximately equivalent to the Duplin.

UNDERLYING FORMATIONS

Zone 2 rests unconformably on Zone 1 along Wiccacon Creek and the Chowan River. Along the latter river the contact can be seen near Mt. Pleasant Landing. Along the lower Neuse, it rests unconformably on the Castle Hayne near Biddles Landing and Fort Barnwell and on the Trent between Rock Landing and New Bern.

OVERLYING FORMATIONS

Where exposed, the Yorktown is almost always covered with thin deposits of Pleistocene terrace materials. The varying thickness of this overburden is one of the factors that determines the value of the Yorktown shell marl for agricultural purposes. Near the coast the Yorktown is overlain by Pliocene, but this relationship can only be determined from well records. A well at Lake Landing in Hyde County shows this relationship.

LITHOLOGY

The material of this upper zone consists chiefly of sands and shell marls with slight admixtures of clay. The marls of the Yorktown (all zones) are rich in lime and have been dug for agricultural purposes at numerous localities.

THICKNESS

As stated above, it is difficult to determine the thickness of the individual zones of the Yorktown formation. The thickness of the entire Yorktown has been estimated from various wells as follows:

	Feet
Atlantic, N. C.	230
Cherry Point, N. C.	140
Cape Hatteras, N. C.	285
Edenton, N. C.	220
Havelock, N. C.	160
Elizabeth City, N. C.	185
Yorktown, Va.	145

According to Miller (1912: 231): "In Hertford, Bertie, Martin, Beaufort and Craven counties it [Yorktown] is rather thin, probably never more than 50 feet in thickness. It may thicken eastward, as the Coastal Plain formations so frequently do, though we have no evidence that such is the case in regard to this formation." Miller was writing of what is now designated as Zone 2, so the thickness of the entire Yorktown may be somewhat greater.

DETAILED SECTIONS

Halifax County

1. *Heathville.*—The farthest inland that Yorktown fossils have been found in North Carolina is a road cut on highway 561 between Heathville and Nevills Store. At this locality some large specimens of *Pecten* were found by M. J. Mundorff.

2. *Halifax.*—Fossils were found in a ravine on a branch of Quankey Creek on the farm of Alex Ponton. Poorly preserved mollusks and whale bones were collected.

3. *Halifax.*—Mr. M. J. Mundorff recently obtained some poorly preserved Miocene fossils from a canyon branching off Roanoke River ½ mile north of the center of Halifax and just south of the County School Bus Shop on highway 301.

4. *Scotland Neck.*—In the North Carolina State Museum there are some whale vertebra collected by L. R. Mills about 4 miles east of Scotland Neck.

5. *Enfield.*—Marl has been reported along Fishing Creek in the vicinity of Enfield. Bones of whale, shark and horse have also been found in this area. No active marl pits are known at the present time, although old pits were reported on the farm of J. H. Sherrod (Vol. 3).

6. *Palmyra Bluff.*—On the right (south) bank of Roanoke River, about 1 mile east of Palmyra, there is an excellent exposure of Yorktown lying on the Tuscaloosa (fig. 32). The following section was noted:

	Feet
Pleistocene:	
Surface clay loam, grading downward into a buff to yellow sandy clay	7
Loose interbedded white and yellow cross-bedded sands, containing numerous clay laminae toward the base in certain parts of the section	12
Compact drab clay, containing fragmental plant remains	3
Gravel layer with occasional cobbles 6 inches in diameter	½
Miocene (Yorktown):	
Fine, loose white, gray, buff, and greenish-gray sands, blotched with iron stains	33
Blue argillaceous sand	2½
Blue argillaceous sand filled with fossil shells and containing many bones, especially near the base	7
Cretaceous:	
Blue argillaceous sand without fossils	2

Northampton County

7. *Near Severn.*—At the bridge over the Meherrin River on highway 35 between Boykins, Va., and Severn, N. C., there is an exposure of fossiliferous blue clay.

Miller (1912) reports similar exposures along the Meherrin River to a point 12 miles below Emporia, Southampton County, Virginia.

8. *Watson's Mills.*—Also known as Worrell's Mills, on Kirby's Creek, about 2.5 miles northwest of Murfreesboro. The following section was noted:

	Feet
Pleistocene:	
Thin bedded drab to yellow sandy clay	10
Miocene (Yorktown):	
Reddish brown sand containing fossil casts	4
Dark blue gray sandy clay with many fossils	16

Whale bones were abundant from the lower layer. *Mulinia congesta* was especially abundant in the upper Miocene strata (fig. 36).

Hertford County

9. *Murfreesboro.*—This is the type locality of the "Murfreesboro" stage of Olsson and also the type locality of several species described by Mansfield. Fossils have been collected at a number of places along the Meherrin River in this vicinity. Miller records them 1½ miles above, 1 mile above, and 1½ miles below the

town. The present survey party obtained a few poor shells from the south bank of the river at the main bridge. The species were not as plentiful as those from Watsons Mill.

Whale bones have been found on the Thompson farm, near Murfreesboro.

10. *Winton.*—The Meherrin River bluffs at Winton consist largely of Pleistocene sands. However, marl has been reported from several farms in the immediate vicinity—for example the G. T. Darden Farm, 6 miles northwest of Winton where shell marl with 78 per cent $CaCo_3$ was reported (Berry and Cushman, 1921: 100).

11. *Mount Pleasant Landing.*—Along the right bank of the Chowan River ¾ mile north of Mount Pleasant Landing, near Harrellsville, there is a layer of blue clay carrying especially *Mulinia congesta* overlain by a layer containing *Venus.* It is thought that this is the slightly unconformable contact between zones 1 and 2 of the Yorktown formation.

12. *Tar Landing Ferry.*—Along the road leading to this small ferry across Wiccacon Creek, 2 miles west of Harrellsville, there is an excellent fossil exposure. The following section is taken from Miller (Vol. 3: 233).

	Feet
Pleistocene:	
Yellow-brown clay loam	13
Miocene (Yorktown):	
Fine gray sand	4
Marl layer, perfectly preserved shells in matrix of broken shells and sand. Sand is blue when unweathered, yellow where exposed. Probably extends to water line	20

About 60 species have been found at this locality. Mansfield (1936: 171) regards this as one of the latest Miocene deposits in northern North Carolina (fig. 30).

Martin County

13. *Hamilton Bluff.*—At the landing on the right (south) bank of the Roanoke River, there is an exposure of sandy clay containing many small shells, especially *Nucula, Leda,* and *Ensis.*

14. *Poplar Point Landing.*—About ½ mile below this landing on the south side of Roanoke River, Mansfield collected some fossils, notably *Pecten eboreus bertiensis* Mansfield.

15. *Abbitts Mill.*—At the bridge of highway 125 over Beaver Dam Creek, a branch of the Roanoke River, there is a deposit of sandy clay about 4 feet thick containing numerous fossils. This locality is 4.8 miles northwest of Williamston.

16. *Williamston.*—Miller (1912) reported numerous marl pits in the vicinity of Williamston, but at the present time none are being operated. There was a new highway sand pit from which numerous fossils were collected 3.8 miles northwest of Williamston (on highway 125) and 1 mile toward the Roanoke River. *Glycymeris* and *Pecten eboreus* were especially common.

Pasquotank County

17. *Elizabeth City.*—Miocene mollusks, foraminifera, and diatoms have been found in well samples at Elizabeth City. The thickness of the Miocene is not determined, but evidence from diatoms suggests that it extends to a depth of 330 feet. The Miocene is overlain by about 50 feet of Pliocene and Pleistocene (Henbest, Lohman, and Mansfield, 1939).

Recent wells at the nearby Blimp Base have confirmed the presence of the Miocene at these depths. The Miocene deposits in these wells are probably the Yorktown formation.

Beaufort County

18. *Washington.*—Yorktown fossils and shark's teeth have been dredged from the Pamlico River at Washington. An exposure of similar Yorktown material can be seen on Runyon's Creek, about 2 miles east of Washington.

Berry and Cushman (1921 : 105) recorded a fertilizer plant (Alfred Styron) utilizing Yorktown marl at Washington but this is not operating at the present time.

19. *Chocowinity.*—Miocene marl has been obtained from shallow excavations under the overpass on highway 17. These two Beaufort county localities probably belong to the later phase of the Yorktown (Zone 2).

Bertie County

20. *Colerain Landing.*—Bluffs along the right (south) bank of the Chowan River rise to 45 or 50 feet. Shells are present throughout most of the section. At the water line, *Glycymeris* is especially common, while *Pecten eboreus bertiensis* becomes more common about 4 feet higher. About 20 feet above the river level large shells of *Venus* and *Rangia clathrodonta* are especially common. The *Rangia* indicates that this is a very late phase of the Yorktown (fig. 33).

21. *Mount Gould Landing.*—Similar fossils occur below the bluff along the south side of Chowan River ¾ mile below Mount Gould Landing. No exposure is present at the landing itself, although numerous fossils are found on the beach where they have been washed from some nearby exposure.

22. *Black Rock.*—An eight-foot section of the bluff, rich in fossils, occurs on the west bank of the Chowan River, about 1½ miles above Eden House Point. Layers of partly indurated black sand occur in the bluff, and a large rock, presumably of the same material, projects just above the river about 100 yards from shore. It is thought probable that this sand-rock is the source of the ilmenite (iron-titanium oxide) sand so abundant on the bottom of the Chowan River and Albemarle Sound in this vicinity (fig. 31).

Nash County

23. *Rocky Mount.*—Numerous marl pits were listed by Miller in the vicinity of Rocky Mount. None of these are being worked at the present time and it is difficult to locate the original pits because they have been filled in, or are now filled with water.

An abandoned pit was observed near the Power Plant, just west of Rocky Mount. *Mulinia congesta* was especially abundant in the banks of the pit.

Recently (1943) Samuel W. Hunter of Rocky Mount found some Yorktown fossils near that town.

Edgecombe County

24. *Rocky Mount.*—Since the county line passes through the middle of Rocky Mount, some localities referred to this town are in Nash County and others in Edgecombe County.

25. *Four Miles Southwest of Battleboro.*—Sandy shell limestone was exposed in a field on the H. N. Davenport farm on the road between Battleboro and Wrendale.

Numerous other pits were reported in this vicinity, but none are active at present.

26. *Shiloh Mills.*—On the left bank of Tar River, 2 miles northwest of Tarboro along highway 44 at the bend in the river, the following section was noted:

	Feet
Pleistocene:	
Brown sandy loam, grading downward into brown sand containing numerous very small pebbles, sand distinctly laminated	10
Miocene (St. Mary's) = Yorktown Zone 1:	
Blue argillaceous fine sand	3½
Shell bed; perfect shells in matrix of blue sand, grading downward into lower member, a few well-rounded quartz pebbles as much as 1 inch in diameter	3½
Fine blue sand, weathering to gray in color, containing a few shells, grading into member below	4
Shell bed, perfect shells containing species similar to shell bed above	
Near base there is a well-defined layer of *Ostrea sculpturata*	
Also, in lower part extending up to 1–2 feet are a great many well-rounded quartz pebbles, some more than an inch in diameter; pieces of lignite, 1 piece 2½ feet long and 6 inches in diameter; fragments of large bones, moderately water-worn, and large sharks' teeth	3–5
Undulating contact	

27. *Bell's Bridge.*—At the site of Old Bell's Bridge on Tar River, about 3 miles northwest of Tarboro, there is a low bluff containing shells at the water line. *Divarcella quadrisulcata* is especially common.

28. *Tarboro.*—Similar exposures are reported by Miller at other places along the Tar River near Tarboro. In addition, the presence of former marl pits in this area showed the presence of Yorktown deposits at shallow depths.

29. Old Sparta.—A 60-foot bluff occurs in a tributary ravine of Tar River, 1 mile south of Old Sparta. The section is as follows:

	Feet
Sandy loam ....Pleistocene	15
Clay with few fossils ... Miocene (Yorktown)	27
Sandy clay, no fossils ...Cretaceous (Tuscaloosa) ...	15

The fossils are very poorly preserved.

30. Leggett.—At Bryant's Bridge over Fishing Creek 3½ miles northeast of Leggett, a fossil exposure was noted on the right bank of the stream about 150 yards below the bridge. A *Turritella* layer was noted above a *Pecten* layer. Fossils occurred up to 5 feet above normal river level.

A similar exposure was recorded by Miller (p. 214) along Fishing Creek ⅓ mile below Mabrey's bridge, near Leggett.

Wilson County

31. Wilson.—Abandoned marl pits occur in the vicinity of Wilson, especially along Toisnot Creek and Hominy Swamp, but none are accessible at the present time.

32. Stantonsburg.—Excavations for a Sewage Disposal Plant in 1939 uncovered numerous whale bones and some sharks' teeth.

Pitt County

33. Farmville.—Miocene fossils have been reported from canal banks near Farmville. At the time of our survey the only fossils observed came from the banks of Little Contentnea Creek on highway 25 between Farmville and Fountain. The shells were at the bottom of the creek.

Whale bones from the vicinity of Farmville are in the North Carolina State Museum.

34. Greenville.—Miller records Miocene fossils from many pits in the vicinity of Greenville. Most of these are now abandoned. However, good Yorktown fossils were obtained from (*a*) a creek on the east side of East Carolina Teachers' College campus and (*b*) a road ditch 2 miles northwest of Greenville on highway 43.

35. Conetoe Swamp.—Dredgings for the canals draining Conetoe Swamp passed through Pleistocene (Wicomico), Miocene (Yorktown), and Cretaceous (Black Creek) formations. The best Miocene fossils came from a place about 5 miles west southwest of Bethel.

36. Five Miles South of Ayden.—Marl pits belonging to J. P. Dawson, 5 miles south of Ayden and 1 mile west of highway 11, formerly yielded a rich shell marl (93 per cent lime) overlying a blue clay. The pit has not been dug recently, although numerous shells (particularly *Mulinia congesta*) were collected by the side of the pits.

Miller (pp. 227–228) recorded old marl pits near Tugwell, and Standard, but none of these are accessible at the present time.

Greene County

37. Maury.—An extensive marl pit is now being operated by C. L. Hardy, 2 miles east of Maury and ¼ mile south of highway 102. Perfect shells were very abundant. Here the Miocene overlies the Snow Hill phase of the Black Creek (Cretaceous) as shown by the few Cretaceous shells in the pit. The Frizzell Pits adjoin those of Mr. Hardy (fig. 35).

38. Lizzie.—Similar marl pits formerly were operating at Lizzie and Castoria. A few shells were obtained near the former settlement.

Wayne County

39. Pikeville.—Drainage excavations in Nahunta Swamp near Pikeville have yielded numerous invertebrates as well as some whale bones (1930).

40. Walters.—Miller recorded marl pits 1¼ miles west of Walters.

Craven County

41. Maple Cypress.—Here the Yorktown marl is a thin mantle on top of the Castle Hayne. It is possible that the Trent limestone occurs at this locality also, but this latter identification is uncertain. A visit to some pits ½ mile south of Maple Cypress uncovered some whale bones lying on top of the Eocene shell marl. These are probably of Yorktown age.

42. Cannon's Marl Pit.—Yorktown marl unconformably overlies the more consolidated Castle Hayne at this pit which is located 2 miles east of Fort Barnwell on the right bank of the Neuse River. Loose shells have been dug extensively for agricultural purposes and some 25 species have been identified. The upper surface of the later Eocene rock has been extensively bored by Yorktown mollusks, especially *Xylophaga* sp. Apparently the Castle Hayne rock was exposed as a ledge as the Yorktown seas advanced over the region.

43. Fort Barnwell.—Marl pits on the property of Z. B. Broadway, 1 mile north of Fort Barnwell contain loose Yorktown marl on top of consolidated Castle Hayne limestone. The Miocene is overlain by a muck deposit of Pleistocene age containing remains of Mastodon, Tapir, etc. (see page 42).

44. Rock Landing.—Old marl pits on the John Daugherty property, at Rock Landing on the Neuse River, 2 miles west of Jasper and 16 miles above New Bern contain some typical Yorktown fossils. In former days these pits were much more extensive and large collections of these fossils were obtained by Miller and other earlier workers.

45. Camp Battle.—Barnacles, probably of Yorktown age, are attached to the Trent limestone which is exposed along the right bank of the Neuse River, 3 miles above New Bern and ¼ mile above the Pumping Station (see page 19).

Hyde County

46. *Lake Landing.*—Yorktown fossils were reported by Miller (p. 253) between the depths of 100 and 200 feet in a well at Lake Landing. These were overlain successively by deposits of Pliocene (Waccamaw) and Pleistocene (Pamlico) age. A summary of the well log is as follows:

	Feet
Pleistocene (Pamlico)	0– 80
Pliocene (Waccamaw)	80–100
Miocene (Yorktown)	100–200

CORRELATION

The Yorktown formation of North Carolina correlates in general with the formation of the same name in Virginia. The Virginia deposits are somewhat better known than those of North Carolina and have been divided into several zones. The following table is based upon Mansfield's study of the family Pectinidae (Mansfield, 1936a: 173):

TABLE 5
CORRELATION OF YORKTOWN FORMATION OF VIRGINIA AND NORTH CAROLINA

VIRGINIA			NORTH CAROLINA	
YORKTOWN FORMATION	ZONE 2	Beds at Suffolk	Uppermost Yorktown	Duplin marl
		Beds at Yorktown	Equivalent of beds at Yorktown, Va.	
		Chama-bearing bed	Equivalent of *Chama*-bearing bed	
		ZONE 1	ZONE 2	

According to Mansfield:

The Yorktown [late Miocene] in the northern part of the State [North Carolina] contains a fauna suggestive of deposition in colder water than that of the beds in the southern part of the State. The cold-water fauna apparently lived in an embayment whose waters transgressed the older rocks of the Piedmont plateau; the fauna, especially that near the inner shore line, was consequently somewhat protected from the influence of the warmer oceanic waters that lay east of the embayment.

The bed at Murfreesboro (Zone 1) forms the lower bed of the Yorktown and is overlain by a bed corresponding to the *Chama* bed of the James River, Virginia. This same bed is found near Halifax, N. C. It is quite possible that many of the deposits referred by Miller (Vol. 3) to the St. Mary's and by Olsson to the "Murfreesboro" stage, and in this report to Zone 1 of the Yorktown, may actually belong to the basal part of Zone 2 as defined by Mansfield. Detailed work on the fossils will be necessary before the exact subdivisions of the Yorktown in North Carolina can be determined.

According to Mansfield (1936a: 171) the latest late Miocene fauna in the northern part of North Carolina is exposed at Tar Ferry, Wiccacon Creek. He believes that it is comparable in age to the Duplin marl at Natural Well and a little younger than the Yorktown that is exposed in the vicinity of Suffolk, Virginia.

MacNeil (1938: 19) has prepared a correlation table (table 6) of the late Yorktown formations after a study of the subfamily Noetinae (Arcidae).

An excellent discussion of the fauna of the Yorktown was given by Dr. Julia Gardner in extended notes quoted by Clark in Volume 3. It should be noted, however, that the correlations used by Dr. Gardner were those current at that time in which the lower part of the Yorktown was regarded as St. Mary's. A further discussion is given in her recent publication (Gardner, 1943).

TABLE 6
CORRELATION OF DUPLIN FORMATION

	SOUTH CAROLINA	NORTH CAROLINA					VIRGINIA
		Southern	S. Central		N. Central	Northern	
DUPLIN MARL	?	Beds around Lumberton		YORKTOWN FORMATION		Beds in vicinity of Chowan River	
	Beds in vicinity of Darlington and Mayesville				Highest bed in Greene Co.	?	
			Beds in Duplin and Sampson Co.		Pitt, Greene, Craven Cos.		Franklin
							?
							Beds at Suffolk

FOSSILS FROM YORKTOWN FORMATION

For further lists and figures see Miller *et al.* (Vol. 3) and papers by Gardner (1943) (1948).

Gastropoda

Crepidula fornicata Linne 2, 5, 6, 8, 20, 22, 34, 37
C. aculeata Gmelin 6, 22
C. plana Say 22, 37
Crucibulum costatum Say 8, 37
Calliostoma conradinum Dall 6, 8, 37
Conus adversarius Conrad 5
Ecphora quadrisulcata Say 34
Fissurella redimiculata Say 5, 8, 37
Fulgur canaliculata Say 12
F. perversum Linne 5
F. maximum Conrad 22
Lirosoma sulcata Conrad 8
Marginella limatula Conrad 15, 20, 22
Murex conradi Dall 22
Natica interna Say 12, 37
Oliva sayana Ravenel 5
Olivella mutica Say 12, 20, 22, 37
Polinices heros Say 6, 12, 34
P. duplicata Say 5, 20, 22
Scaphella obtusa Emmons 37
Siphonalia migrans Conrad 37
Terebra unilineata Conrad 37
T. simplex Conrad 37
T. carolinensis Emmons
Turritella variabilis Conrad 5, 6, 12, 20, 22, 23, 26, 27, 30, 34
Urosalpinx cinerea Say 2, 20, 22, 34
U. trossulus Conrad 8
Vermetus graniferus Say 5, 12, 22, 37
V. sculpturata Lea 18, 37

Pelecypoda

Arca improcera Conrad 37, 43
A. carolinense Conrad 12, 16, 20, 22
A. incile Say 12, 18, 20, 21, 22, 37
A. limula Conrad 20
Astarte concentrica Conrad 12, 18, 20, 22, 26, 27, 35, 37
A. undulata Say 6, 12, 18, 27, 34, 35
A. bella Conrad 43
A. symmetrica Conrad 6, 18, 20, 35
Anomia simplex D'Orbigny 20, 26, 30, 37
Cardium sublineolatum Conrad 43
Cardita arata Conrad 20, 43
Chione latilirata Conrad 22
Chama congregata Conrad 27
Corbula inaequalis Say 6, 12, 20, 22, 26, 27, 34
C. cuneata Say 12
Crassatellites undulatus Say 6, 12, 16, 22, 26, 27, 35, 37, 43
C. gibbesii Tuomey & Holmes 18
Diplodonta acclinis Conrad 22
Divarcella quadrisulcata D'Orbigny 20, 22, 27

Ensis directus Conrad 6, 12, 18, 20, 22, 27, 37
Glycymeris pectinata Gmelin 20, 22, 26, 37
G. subovata Say 12, 18, 20, 27, 34, 35, 37, 43
G. americana De France 20
Leda acuta Conrad 12, 15, 20, 22, 24
Modiolus ducatelli Conrad 12
Mulinia congesta Conrad 6, 12, 15, 18, 20, 22, 23, 26, 27, 31, 34, 36, 37, 43
Nucula proxima Say 12, 15, 20, 22, 30, 34
Ostrea compressirostra Say 20, 27, 29, 34, 35
O. sculpturata Conrad 6, 18, 20, 26, 27, 29, 35, 37, 43
O. trigonalis Conrad 16, 37
Pecten eboreus Conrad 6, 12, 18, 20, 22, 23, 27, 34, 35, 37, 43
P. madisonius Say 18, 27
P. jeffersonius Say 18, 43
Phacoides crenulata Conrad 12, 15, 18, 20, 22, 27, 34
P. anodonta Say 21, 22, 26, 27
Plicatula marginata Say 18, 20, 37
Rangia clathrodonta Conrad 20
Teredo fistula Lea 6
Venus mercenaria Linne 12, 18, 34
V. rileyi Conrad 12, 35, 37
Venericardia tridentata Say 6, 12, 20, 22, 27
V. granulata Say 6, 12, 18, 20, 21, 22, 26, 34, 35, 37, 43
Yoldia laevis Say 15

Scaphopoda

Dentalium attenuatum Say 6, 20, 22, 26, 27, 30, 36
D. carolinense Conrad 37
Cadulus thallus Conrad 37

Crustacea

Balanus concavus Bronn 5, 18, 34, 35

DUPLIN FORMATION

NAME

Although the rich molluscan fauna at the Natural Well was known many years earlier, the first use of the term Duplin as a formation was by Dall in 1895. It is named from Duplin County, North Carolina, where the beds are best exposed. Although no type locality was given, the Natural Well near Magnolia is the locality from which Dall described most of his Duplin fossils.

EXTENT

Duplin marl occurs locally in the southern part of the State. It is best developed near Magnolia, Duplin County, but is also present in Sampson County near Clinton, in Bladen County near Elizabethtown, in Robeson County near Lumberton and Fairmont, and in Columbus County at Lake Waccamaw. Duplin fossils have also been reported from excavations near Wilmington in New Hanover County.

The Duplin marl extends into South Carolina where it also occurs locally. The best collection of fossils in that State has come from marl pits near Mayesville, Sumter County, S. C. (Cooke, 1936).

UNDERLYING FORMATIONS

In Duplin County, as at the Natural Well, Duplin marl rests unconformably on the Castle Hayne (Eocene). In Sampson County it probably rests on the Cretaceous. Near Wilmington, Duplin marl was reported lying on the Castle Hayne. No Eocene strata have been observed in Bladen, Columbus, and Robeson Counties, and in those areas the Duplin apparently rests directly on the Cretaceous.

OVERLYING FORMATIONS

The Duplin marl is usually overlain by thin deposits of Pleistocene terrace sands. However, along the Cape Fear River (between Wilmington and Elizabethtown) and along the south shore of Lake Waccamaw, it is directly overlain by a thin layer of fossiliferous Waccamaw (Pliocene) shell marl.

THICKNESS

Miller (1912: 239) believed that the Duplin marl was nearly 100 feet thick at the Natural Well. However, Huddle (1940) has shown that the lower part of the section contained Eocene microfossils and therefore the Duplin marl is only about 3 feet thick. Where it rests on the Cretaceous, the Duplin marl is relatively thin (less than 50 feet).

DETAILED SECTIONS

Duplin County

1. *Natural Well.*—Two miles southwest of Magnolia is the famous Natural Well from which so many Miocene fossils have been described. It is a circular sink hole, 75 to 100 feet in diameter and about 30 feet deep. One wall is very steep but on one side it is somewhat sloping so that it is possible to reach the bottom where there is a pond, about 30 feet in diameter. The section is as follows:

		Feet
Sand	Pleistocene	10
White marl, many shells, sand	Duplin	3
Green sandstone, lightly cemented	Castle Hayne	9

Miller regarded the two lower beds as Duplin; however, Huddle has shown that the lower bed contained some Eocene microfossils.

The fauna from the Natural Well is exceedingly rich, something over 400 species having been reported from there. The most complete list is given by Miller (Vol. 3: 241–244); additional records have been added by Johnson (1904), Olsson (1914, 1916), Gardner (1943, 1948), etc. (fig. 38).

Other deposits of Miocene (Duplin) shell marl have been reported from the immediate vicinity of Magnolia, but nowhere else is there the rich fauna of the Natural Well.

Pender County

2. *Watha.*—Marl was formerly dug on the farm of A. A. McMillan on Lewis Creek, 1 mile south of Watha.

Six to 8 feet of sand overlie the Duplin marl, which is here about 2½ feet thick. It overlies the Castle Hayne (Eocene).

3. *St. Helena.*—A well at St. Helena, 2½ miles south of Burgaw showed the Duplin formation between the depths of 20 and 60 feet. It overlies the Peedee (Cretaceous).

Bladen County

4. *Tar Heel.*—Marl pits on the property of W. R. Robeson near the Cape Fear River between Kings Landing and Tar Heel Landing have yielded a large collection of fossils. The majority of the specimens are of Duplin age and are similar to those from the Natural Well. It is believed that the upper part of the section belongs to the Waccamaw (Pliocene) formation, as indicated by the presence of such species as *Rangia clathrodonta*.

A similar pit occurs on the farm of J. E. Robeson about 1 mile up river.

5. *Elizabethtown.*—Miller records Duplin marl 1½ and 4 miles south of Elizabethtown. However, material recently obtained from a road cut near Browns Creek, 1½ miles south of town, seemed to belong to the Waccamaw formation (see page 32).

6. *Clarkton.*—Old marl pits were reported by Miller (p. 248) 4 miles south of Clarkton on Brick Yard Branch.

Sampson County

7. *Clinton.*—Miller (p. 240) recorded old marl pits 4 miles south of Clinton and also stated that outcrops occurred on Duncan Branch and Gum Chimney Branch, 2½ miles south and 2½ and 3 miles south of Clinton.

According to Berry and Cushman (1921: 127) Duplin marl occurs in Sampson County from the headwaters of Six Runs to the vicinity of Lissa and Taylors Bridge and westward beyond Clinton to Great Coharie Creek.

Whale bones have been reported from "marl beds" in this County.

Columbus County

8. *Lake Waccamaw.*—A more or less consolidated shell layer of Duplin age is exposed along the north shore of Lake Waccamaw from the Pumping Station eastward (fig. 37). The following section was noted:

		Feet
Loam, sand and clay	Pleistocene	
Loose shell marl in sandy clay	Waccamaw	2–6
Compact limestone with casts of fossils especially *Credidula fornicata*	Duplin	3
Sand with phosphatic pebbles and casts of Cretaceous from Peedee fossils reworked to waterline	Duplin	1
BELOW WATER ————	Peedee	

9. *Whiteville.*—Miller (p. 248) recorded Duplin marl in old pits near Whiteville.

New Hanover County

10. *Wilmington.*—Miller and other earlier workers recorded Miocene (Duplin) marl from excavations near Wilmington. A restudy of some of this material by Mansfield (1936: 668) has shown that it should be referred to the Pliocene (Waccamaw). This applies to material from (*a*) sewer at corner of Nut and Mulberry Streets, collected by T. W. Stanton in 1891 (U.S.G.S. 2295) and (*b*) near Y.M.C.A. building, collected by T. W. Vaughan in 1902 (U.S.G.S. 3610).

On the other hand a few specimens collected by T. A. Conrad, now in the United States National Museum (6112), labeled as coming from Wilmington, are *Crucibulum ramosum* Conrad, *Glycymeris subovata* Say, and *Panope reflexa* Say, which according to Mansfield are definitely Miocene species. So, apparently both Duplin and Waccamaw marl occur locally in the Wilmington area.

At the old City Quarry near Smith's Creek in the east side of Wilmington, some of the Eocene rock was collected on which had been attached specimens of *Pecten, Plicatula*. Since these could not be specifically identified it was impossible to determine whether they were of Miocene or Pliocene age.

Robeson County

11. *Lumberton.*—Several old marl pits were reported by Miller and Berry and Cushman in the vicinity of Lumberton. On the present survey, material was obtained from two localities: (*a*) Left bank of Lumber River, 2 miles east of Lumberton. Here the shell marl is exposed along the river and has also been dug along the highway (211) nearby; (*b*) 4 miles north of Lumberton on route 301; a few shells were obtained from abandoned pits.

Dickson McLean, Jr., has presented the North Carolina State Museum with a number of sharks' teeth which he obtained from the banks of the Lumber River (1938).

12. *Fairmont.*—Marl pits were formerly dug along branches of Hog Swamp and Old Field Swamp near Fairmont, but none are active at the present time.

Old pits were also reported near Barnesville, Orum, and Ashpole.

Cooke (1936) on his geological map of South Carolina shows Duplin marl near Hamer, Dillon County, South Carolina, adjoining Robeson County, but local inquiry revealed that the shell marl in that vicinity was deeply buried (about 15 feet).

CORRELATION

The fossils from the Duplin marl are probably equivalent in age to those of the upper part of the Yorktown formation but suggest deposition in warmer water (see table on page 54.

The extensive fauna of the Natural Well is closely related to material from marl pits near Mayesville, S. C. (Gardner and Aldrich, 1919).

FOSSILS FROM DUPLIN FORMATION

For further lists see Volume 3 (1912) and Gardner (1943 and 1948).

Gastropoda

Conus adversarius Conrad 1, 4
C. floridanus Gabb 4
Crepidula fornicata Linné
C. plana Say 4
C. spinosum Sowerby 1
Crucibulum constrictum Conrad 4
C. costatum Say 1, 4
C. grandis Say 4
Cypraea carolinensis Dall 1, 4
Ecphora quadricostata Say 1, 4
Fasciolaria rhomboidea Rogers 1
F. sparrowi Emmons 1
F. apicina Dall 4
F. elegans Emmons 4
Fulgur maximum Conrad 4
F. coronatum Conrad 1
F. perversum Linne 1, 4
Fusus aequalis Emmons 1, 4
Marginella antiquata Redfield 1, 4
M. limatula Conrad 1, 4
Mitra carolinensis Dall 4
Natica interna Say 1, 5, 8
Oliva sayana Ravenel 1, 4, 5
Olivella mutica Say 1, 4, 8
Polinices duplicata Say 1, 4, 8
P. heros Say 4
Scaphella obtusa Emmons 4
Sinum perspectivum Say 4
Terebra dislocata Say 4
T. carolinensis Conrad 5
Trochus humilis Conrad 4
Turritella variabilis Conrad 1, 4, 5
T. subanulata Heilprin 4, 8
Vermetus sculpturata Lea 1, 4
V. graniferus Say 1, 4, 5, 8

Pelecypoda

Anomia simplex D'Orbigny 4
Arca plicatula Conrad 1
A. improcera Conrad 1, 4
A. carolinensis Conrad 4
A. lineosa Say 4, 11
A scalaris Conrad 4
A. triginitaria Conrad 4, 8
A. limula Conrad 8
A. variabilis MacNeil 8
A. secticostata Reeve 4
Astarte bella Conrad 1
A. obruta Conrad 4
A. undulata Say 1
A. concentrica Conrad 4, 8, 11
Cardium sublineolatum Conrad 1, 4
C. muricatum Linné 4

Chama striata Emmons 1
Chama corticosa Conrad 4, 8
C. arcinella Linné 4
Chione latilirata Conrad 1, 4, 8
Corbula inaequalis Say 8, 11
Crassatellites gibbesii T. & H. 11
C. undulatus Say 1, 4
Divarcella quadrisulcata D'Orbigny 4, 11
Ensis directus Conrad 1, 4, 8
Glycymeris americana DeFrance 1, 4, 11
G. subovata Say 1, 4, 8, 11
G. pectinata Gmelin 8, 11
Leda acuta Conrad 8, 11
Modiolus ducatelli Conrad 8
Mulinia congesta Conrad 4, 8, 11
Mytilus conradinus D'Orbigny 1
Ostrea compressirostra Say 4, 8
O. sculpturata Conrad 8
Pecten eboreus Conrad 1, 8
Phacoides anodonta Conrad 4
P. crenulatus Conrad 4
P. trisulcatus Conrad 4, 11
Plicatula marginata Say 1, 4, 11
Rangia clathrodonta Conrad 4
Tellina declivis Say 4
Venus campechiensis Gmelin 1
V. rileyi Conrad 4, 8
V. tridachnoides Lamarch 4
V. mercenaria Linné 8
Venericardia tridentata Say 1
V. granulata Say 1, 4, 8, 11

Note: Localities 4, 5, and 8 contain a mixture of Duplin and Waccamaw fossils.

PLIOCENE
WACCAMAW FORMATION
NAME

Named by Dall (1892: 209) from the Waccamaw River in South Carolina.

EXTENT

In North Carolina the Waccamaw formation outcrops only south of the Neuse River and is best exposed at several places along the Cape Fear River (Neill's Eddy Landing Acme, Walker's Bluff, and Tar Heel). It also occurs in the upper layer at Lake Waccamaw, Columbus County, and has been dredged from the inland waterways at Stump Sound, Onslow County.

The Croatan beds in the vicinity of the Neuse River may be contemporaneous with the Waccamaw or may be slightly younger (see next chapter). Waccamaw fossils have been reported from a well at Lake Landing in Hyde County, N. C., and shells of questionable Pliocene age have been dredged from the Dismal Swamp Canal near South Mills, Camden County, N. C. These are the farthest north that any marine Pliocene

has been found along the Atlantic seaboard and it is thought that the Pliocene shore line crossed the present shore line somewhere between Cape Hatteras and Cape Henry.

The typical fauna of the Waccamaw formation is well exposed along the Waccamaw River in Horry County, South Carolina, and has been dredged from the Inland Waterways near Little River and Myrtle Beach in the same county.

UNDERLYING FORMATIONS

Along the Cape Fear River the Waccamaw rests either on the Peedee (Cretaceous) or on an intermediate layer of Duplin (Miocene). At some places (Tar Heel) the two (Waccamaw and Duplin) have been dug together for agricultural purposes. Along the north shore of Lake Waccamaw the Pliocene forms a thin unconsolidated layer on top of a more or less compact phase of the Duplin (Miocene). Along the Waccamaw River the Pliocene lies unconformably on the Peedee.

OVERLYING FORMATIONS

The Waccamaw formation is overlain by Pleistocene deposits. At most places the Pleistocene is unfossiliferous ("Higher Terraces") but in a few places (Stump Sound, N. C., Little River, S. C., and Myrtle Beach, S. C.) the Waccamaw is overlain by the fossiliferous Pamlico. It is also overlain by the Pamlico formation in wells in Hyde, Carteret, Dare, and Craven Counties.

LITHOLOGY

It resembles the Yorktown in consisting of loose gray to buff fine (quartz) sand with pebbles and shell beds. Phosphatic pebbles are common and have probably been derived from the overlying Cretaceous (Stump Sound, Little River, etc.). Shell marls are locally of economic importance.

THICKNESS

In North Carolina the Waccamaw is thin, rarely exceeding 25 feet in thickness. In the Hyde County well, it is 20 feet thick.

DETAILED SECTIONS
Onslow County

1. Stump Sound.—At Tar Landing on Stump Sound, southeast of Folkstone, excavations for the Inter-Coastal Canal uncovered a mixture of Pliocene (Waccamaw) and Pleistocene (Pamlico) fossils. Typical Pliocene species from the spoil banks were: Arca subsinuata Conrad, Plicatula marginata Say, Chama congregata Conrad, Phacoides anodonta Conrad, Crassatellites gibbesii, Rangia clathrodonta Conrad, Mellita caroliniana.

Columbus County

2. Neil's Eddy Landing.—On the right bank of the Cape Fear River, 3 miles northeast of Acme is a fa-

mous fossil locality. Formerly there were marl pits at this place belonging to B. F. Keith, but these are not being worked at the present time. A few specimens were obtained from the abandoned pits.

At several places along the right bank of the river fossiliferous (Waccamaw) Pliocene deposits can be seen overlying the (Peedee) Cretaceous clays, but in most places they are practically inaccessible. However, at one spot about ⅛ mile below the former landing the top layer had slumped and it was very easy to obtain a good collection of fossils. This spot is very close to the Brunswick County line, if not actually over in Brunswick County.

Miller listed 83 species from this locality and stated that the collection had only partly been studied. Dr. Julia Gardner has expanded this list in an unpublished manuscript and the present survey party added a number of species to the already long list.

This is the type locality of *Eucrasatella mansfieldi* MacNeil (fig. 20).

3. *Acme.*—Large collections of Pliocene fossils have been obtained from the pits of the Acme Fertilizer Company, opposite the farm of J. W. Butler. Although these pits have not been extensively used for many years, it is still possible to obtain good fossils.

This is the type locality of *Fasciolaria papillosa acmensis* Smith (1940), and an unusually perfect specimen of this giant shell (12½ inches long) has been presented to the North Carolina State Museum by Mr. Butler.

Bones of the Horse (*Equus complicatus*) Leidy were also dug from this marl pit (Berry, 1931).

The Pliocene is relatively thin and the deeper parts of the excavations passed into the Cretaceous as can be shown by the presence of some shells of *Exogyra costata.*

The former name of Acme was Cronley and as such this locality sometimes appears in the literature.

4. *Lake Waccamaw.*—The upper layer along the north side of Lake Waccamaw belongs to the Waccamaw formation unconformably overlying the Duplin marl (see section on page 29) (fig. 37).

Brunswick County

5. *Below Neil's Eddy Landing.*—The Pliocene (Waccamaw) beds along the right bank of the Cape Fear River extend from Neil's Eddy Landing southward into Brunswick County (see page 10).

Bladen County

6. *Walker's Bluff.*—This is a high bluff (80 feet) along the right bank of the Cape Fear River 13 miles south of Elizabethtown by river and 8 miles by road, which is now on the property of Mr. Monrow. The following section has been recorded:

	Feet	Inches
Pleistocene		
Sandy loam	2	
Mottled reddish and yellowish arenaceous clay	1	6
Drab clay interstratified with yellowish to reddish sand	2	
Laminated drab clay, becoming arenaceous and with fine sand partings in lower half; basal 12 inches iron stained	11	
Pliocene (Waccamaw)		
Coarse, loose, orange-colored sand	3	
Calcareous tough clay		6
Shell marl full of fossils, many perfectly preserved. The perfect specimens are in a matrix of fine shell fragments mixed in places with coarse buff sand. Thickens in places through dipping down in pockets or holes in Cretaceous. Gradually thins out and disappears in middle of bluff through being cut out by overlying Pleistocene. Again appears at bend. It contains some phosphate nodules and some sandstone cobbles 6 inches in diameter. The average thickness is about 5–6 feet, maximum	12	
Cretaceous		
Laminated clays and sands, containing Cretaceous fossils in places, exposed to water's edge	48–51	

This locality has yielded Pliocene fossils second only to Neil's Eddy Landing. However, at the present time, collecting is not very good owing to bad slumping of the bluff.

7. *Elizabethtown.*—A road cut along highway 87 near Brown Creek, 1½ miles south of Elizabethtown, shows a small pocket containing typical Waccamaw species lying on the Black Creek clay. This is probably very close to the place mentioned by Miller as Miocene (see page 29) (fig. 39).

8. *Tar Heel.*—The upper part of the marl pits of W. R. Robeson at Tar Heel is probably of Pliocene age.

New Hanover County

9. *Wilmington.*—As stated previously (page 30) many of the localities previously referred to the Duplin are probably actually of Waccamaw age (see Mansfield, 1936a; 668). The few shells on top of the Castle Hayne limestone that was dredged from the City Rock Quarry are also probably of Waccamaw age.

CORRELATION

The Waccamaw beds of the Carolinas have been correlated with the Caloosahatchie formation of Florida and are usually regarded as being of early Pliocene age.

An excellent discussion of the Waccamaw fauna was given by Dr. Gardner in Volume 3 in which she shows the transition between the Duplin (Miocene) and the Waccamaw (Pliocene). She shows that the fauna represents an early Pliocene facies in which a few of the more stable Miocene forms have survived in greatly diminished numbers, while many of the species which are to become prominent in the later Pliocene, such as the *Arca rustica* and *A. scalarina* are only just beginning to make their appearance. While it may seem to be a mixed fauna, this is more obvious in lists than in the collections, for in the former all species are equally prominent while in the latter the forms suggestive of the Miocene and of the middle and upper Pliocene are inconspicuous elements.

The Croatan fossils from the vicinity of the Neuse River may be contemporaneous with those of the Waccamaw formation or they may be slightly younger.

CROATAN FORMATION

NAME

Dall (1892: 209, 213–217) proposed the name Croatan for beds along the estuary of the Neuse River. It was named from the small settlement of Croatan, about 10 miles below New Bern in Craven County, N. C. Dall regarded this as somewhat younger than the Waccamaw.

Mansfield (1928) pointed out that some of Dall's Croatan beds contained mixtures of Pliocene and Pleistocene (Pamlico) fossils. He therefore proposed to restrict the term Croatan to the beds on or near the Neuse River which are of Pliocene age. In 1936 he reported additional Croatan localities in Craven and Onslow Counties. He regarded these as approximately equivalent in age to the Waccamaw.

EXTENT

The Croatan sand is best developed along the south side of the Neuse River between James City and Cherry Point and from excavations for the Intracoastal Waterway canal near Beaufort. It also occurs locally southwest of the Neuse River in Craven and Onslow Counties and has been dug for marl near Kuhns and Silverdale not far from the White Oak River. The farthest from Neuse River that typical Croatan has been reported is near Padgett in Onslow County (Mansfield, 1936, and this report locality 10).

As stated above, it is probable that the Croatan sand is continuous with the Waccamaw, although it is also possible that it may represent a younger phase, as originally suggested by Dall.

UNDERLYING FORMATIONS

Wherever observed, the Croatan sand lies unconformably on the Trent (Miocene) marl. At the Askew marl pit (near Silverdale) the two marls have been dug together for fertilizer.

OVERLYING FORMATIONS

The Croatan is always overlain by the Pleistocene. Along the Neuse River and at Core Creek, near Beaufort, it is covered with the marine Pamlico formation

and the fossils have frequently been mechanically mixed. At one place, on the Neuse, 10 miles below New Bern, the Croatan is separated from the Pamlico from the Horry clay which is thought to represent a low stand of the sea immediately preceding the deposition of the Marine Pamlico (see page 39). At Silverdale and Padgett the Croatan is overlain by nonfossiliferous Pleistocene.

THICKNESS

The maximum thickness is near James City where it is something over 15 feet.

DETAILED SECTIONS

Craven County

1. *James City.*—A good exposure of fossiliferous Croatan sand occurs on the right bank of the Neuse River, 2 miles below James City. The fossiliferous bed is here about 15 feet thick and extends along the river for 100 yards. This locality was listed by Mansfield (1936: 665). At that time it was the property of Mr. Hastings; now it is owned by Mrs. Nettie Farrell. Large corals (*Astrangia danae*) are especially conspicuous (U.S.G.S. 13812) (fig. 40).

2. *Riverdale Wharf.*—9 miles below New Bern. Here the unfossiliferous Pleistocene rests on a thin deposit of Croatan sand. This locality was listed by Miller (p. 253).

3. *Eleven Miles Below New Bern.*—The following section was taken on the Boyd Property:

	Feet
Sandy soil	2
Pleistocene	
Laminated reddish sand and clay with a water seepage at the base	6
Gray clayey sand	8
Very fossiliferous fine-grained sand	4
Unconformity	
Pliocene (Croatan sand)	
Concretionary, ferruginous coarse sand and gravel, carrying corals and mollusks	0–2

Recent erosion along the river has obliterated the lower part of the section and it is now impossible to observe the Pliocene-Pleistocene contact.

4. *Croatan.*—Mansfield (1928) records several localities in the vicinity of Croatan and Riverdale where excavations have passed through both Pleistocene and Pliocene formations. Recent pits visited near Croatan were not deep enough to encounter the Pliocene.

5. *Twelve to Fifteen Miles Below New Bern.*—Mansfield (1928) records additional localities along the Neuse River where he obtained Croatan fossils. Sometimes it was difficult to separate them from the overlying Pleistocene.

Pliocene marine shells which can be referred to either the Waccamaw or Croatan formation have been reported in wells at Cherry Point, Havelock, and elsewhere in the regions. It is usually difficult to recognize the boundary between the Pliocene and the Pleistocene in these wells.

Carteret County

6. *Core Creek.*—Excavations for the Intra-Coastal Canal near Core Creek Bridge apparently passed through a thin layer of Pleistocene into the Croatan. The spoil banks one mile northeast of the bridge contained such Pliocene species as *Arca subsinuata* Conrad, *Cardita arata* Conrad, *Phacoides anodonta* Conrad, and *Pecten eboreus* Conrad.

7. *Kuhns.*—Several marl pits occur near Kuhns, although none are active at the present time. Fossils were obtained from an abandoned pit on the Koonce property.

Canu and Bassler (1923: 9) listed 6 species of bryozoa from Kuhns, of which 3 were new. They recorded the deposit as Miocene, but the mollusks certainly suggest a Pliocene age.

Onslow County

8. *Grants Creek.*—Marl pits carrying typical Croatan fossils occur on the farm of S. G. Jones at the edge of a swamp on Grants Creek, a tributary of Whiteoak River, about 5 miles west of Silverdale. This is approximately opposite the localities at Kuhns in Carteret County and probably is of the same age.

9. *Silverdale.*—The highest fossiliferous bed in the marl pits belonging to A. W. Askew can be referred to the Croatan. The lower part is definitely in the Trent (Miocene) formation and contains fossils similar to those at the Gillette pit. It is impossible to draw a sharp line between the two formations and it seems probable that there has been some mechanical mixing of the two deposits, probably during the Pliocene. The Croatan beds are definitely characterized by the many shells of *Pecten eboreus.*

10. *Padgett.*—Mansfield recorded two marl beds near Padgett. At the time of our visit, none were in operation although we obtained some fossils from an abandoned pit on the property of A. Z. Thompson, 2 miles north of Padgett.

LOCALITIES OF UNDIFFERENTIATED PLIOCENE— WACCAMAW OR CROATAN

Hyde County

11. *Lake Landing.*—According to Miller (Vol. 3: 252–253) this well passed fossiliferous Pliocene between 80 and 100 feet on top of the Miocene and beneath the Pamlico.

12. *Lake Matamuskeet.*—Dredgings for canals have yielded a large number of Pleistocene species. Among this material are a few specimens of apparent Pliocene age. For example there is a shell of *Ostrea compressirostra* Say dredged from Lake Matamuskeet and some specimens of the Coral *Astraea crassa* dredged from the

Intra-Coastal Canal in the same County (Richards, 1936: 1632).

13. *Ocracoke.*—Pliocene and Pleistocene fossils are occasionally washed onto the beach between Hatteras and Ocracoke (Richards, 1936).

Dare County

14. *Hatteras.*—Pliocene shells have occasionally been washed onto the beach (see above). They were also found in the recent well but are difficult to separate from the overlying Pleistocene.

Pasquotank County

15. *Elizabeth City.*—According to Mansfield (Henbest, Lohman, and Mansfield, 1939), Pliocene (Croatan?) sediments were encountered in wells between 50 and 55 feet, separating the Pleistocene and the Miocene.

Camden County

16. *South Mills.*—The fossils dredged from the Dismal Swamp Canal are largely of Pleistocene age. A few such as *Arca subsinuata* Conrad may possibly have come from an underlying layer of Pliocene.

CORRELATION

The fauna of the Croatan formation was originally regarded by Dall as being younger than the Waccamaw. This was based upon the high percentage of Recent forms. However, it was shown by Mansfield (1928) that the Croatan beds of Dall were partly a mixture of Pliocene and Pleistocene deposits. Therefore, Mansfield regarded the two Pliocene deposits as contemporaneous.

The Croatan beds, plus the few Pliocene beds listed from Hyde, Pasquotank, and Camden Counties (localities 11–16), mark the farthest north that any marine Pliocene deposits have been found along the East Coast. It is believed that the Pliocene shore line crossed the present shore somewhere in the vicinity of Cape Henry, Virginia.

The fauna recently described from the Santee-Cooper Canal, in South Carolina (Richards, 1943), may represent a still later phase of the Pliocene, or may be early Pleistocene.

FOSSILS FROM WACCAMAW AND CROATAN FORMATIONS

Numbers preceded by C refer to Croatan formation. See also lists by Miller *et al.* (Vol. 3, 1912), Gardner (1943, 1948).

Gastropoda

Cerithium muscarum Say 3
Calyptraea aperta Solander 1
Conus marylandicus Green 2, 3
Crepidula fornicata Linné, 1, 2, 3, C1, C8, C10
Fasciolaria rhomboidea Rogers 3
F. apicina Dall C1
F. papillosa acmensis Smith 3

Fulgur maximum Conrad 6
F. perversum Linné 3, C8
F. carica Gmelin 1
F. pyrum Conrad 3
Marginella limatula Conrad 1, 3
M. apicina Menke 2, C1, C8, C10
Nassa obsoleta Say 1
N. vibex Say 2
N. trivittata Say 2, C10
Natica interna Say 2
N. canrena Lamarck 2
Oliva sayana Ravenel 1, 2, 3
Olivella mutica Say C7, C8, C10
Polinices duplicata Say 1, 2, 3, 6, C1, C8, C10
Terebra dislocata Say 1, 2, 3
Turritella variabilis Say 2, C1
Urosalpinx cinerea Say 1, 2, 3, C1
Vermetus graniferus Say 2, C1

Pelecypoda

Anomia simplex D'Orgibny 2, 3
Arca adamsi Smith 2
A. improcera Conrad 2, 3, C1
A. transversa Say 6
A. staminea Say 7
A. subsinuata Conrad 1, C1
A. limula Conrad C1
A. carolinensis Conrad C1
A. variabilis MacNeill C1
Astarte concentrica Conrad 3
A. symmetrica Conrad 2
A. undulata Say C1
Chama congregata Conrad 1, 2, 3, 7, C1
Crassatellites gibbesii T & H 1, 2, 3, C10
Cardium sublineolatum Conrad 2, 3
C. isocardium Linné 2
Chione cribaria Conrad 2, 3
C. latilirata Conrad 2, 3
Corbula inaequalis Say 2, C1, C8
Crassinella lunulata Conrad 6
Divarcella quadrisulcata D'Orbigny 3, 6
Donax variabilis Say 6, C8
Ensis directus Conrad 2, 6, C1
Echinochama arcinella Linné 2, 3
Glycymeris pectinatus Gmelin 1, 3
G. americana De France 2, 3
Leda acuta Conrad 2, C1
Mulinia congesta Conrad 1
M. lateralis Say 1, 2, 3, 6, C1, C8
Macrocallista numbosa Solander 3
Nucula proxima Say 1, 2, C1
Ostrea sculpturata Conrad 2, 3, 6, C1
O. compressirostra Say 3
O. virginica Gmelin 3, 6, C1
Plicatula marginata Say 1, 2, 3, 6, C1
Pecten gibbus Linné 1
P. eboreus Conrad 3, C1
Phacoides anodonta Say 1, 6

P. crenulatus Conrad 2, 3
Rangia clathrodonta Conrad C1, C8
Semele proficua Putteney 1
Venus campechiensis Gmelin 1
V. mercenaria Linné 3, 7
V. rileyi Conrad C1, C10
Venericardia granulata Say 2, 3, 6
V. tridentata Say 2

HIGH-LEVEL GRAVELS

This term is used for gravels, sands, and some clay formerly covered in part by the names "Appomatox" and "Lafayette" formations. Under this later name, these gravels were described at some length by Stephenson (Vol. 3). The name "Lafayette," derived from Lafayette County, Mississippi, was first used by Hilgard in 1891. It was later pointed out that the type locality of this formation was actually of Eocene age, and therefore the name could not apply to the Pliocene deposits along the Atlantic Coastal Plain (Berry, 1911).

Mundorff (1946) in discussing the Halifax area of North Carolina speaks of

unclassified high-level gravel, sand and clay . . . mostly of fluviatile origin deposited as stream channel, stream terrace and basin filling. . . . It is found only above 270 feet and occurs in patches and pockets. . . . The age of these deposits is indefinite. . . . They may include representatives of several ages, some possibly as old as the Cretaceous.

No attempt will be made in the present report to define the distribution or to determine the age of these high-level gravels. A few detailed sections of typical exposures are given below.

DETAILED SECTIONS

Nash County

1. *Between Samaria and Stanhope.*—Very irregular patches of gravel resting on the bed rock can be seen along Highway 95 between Samaria and Stanhope. Similar exposures are known elsewhere in Nash County and in practically all cases the gravel occupies very irregular surfaces of the bed rock (fig. 44, 45).

Harnett County

2. *Lillington.*—Gravel lies on top of the Eocene rocks at the abandoned pit of the Cape Fear Gravel Company, 2 miles northwest of Lillington (see page 14). It is impossible to draw the line between the weathered surface of the weathered Eocene and the surface gravel (fig. 43).

3. *Little River.*—Extensive deposits of gravel are being dug from the Bryan Monrow Pit, near the crossing of Little River of highway 15A, 4 miles south of Lillington.

Moore County

4. *Aberdeen.*—There is an extensive sand pit on highway 211 just south of Aberdeen. The sand is

from the Tuscaloosa formation, but locally it is covered by a thin mantle of gravel, probably of Pliocene age (see page 6).

5. *Vass.*—Pliocene (?) gravel was noted near Vass station of the Seaboard Air Line Railroad.

Richmond County

6. *Hamlet.*—Cuts of the S.A.L. RR. just north of Hamlet show the following section (abbreviated from Stephenson):

	Feet
Pliocene? (Lafayette formation):	
Coarse, loose, pebbly sand, gray at top, becoming a dull yellow towards base. Merges with the underlying red sand and gravel layer towards the north end	2–5
Red and mottled, coarse, argillaceous, more or less pebbly sand, with a partially discontinuous band of pebbles and cobbles along base. Iron crusts present along contact in places. The pebbles and cobbles are of all sizes up to 4 or 5 inches and vary in shape from angular to moderately well rounded	1–8
Cretaceous (Patuxent formation = Tuscaloosa)	
Coarse, cross-bedded, argillaceous, arkosic, varicolored sand, with subordinate clay lenses, and in places numerous rolled clay balls	10–20

Anson County

7. *Bonsall Pits, Lilesville.*—Extensive gravel pits are being operated by the W. R. Bonsall Company, 3 miles south of Lilesville. According to a letter from F. J. Cloud, Vice President of the company (fig. 42):

Our deposits are first made up of a wind-blown sand layer which is closely akin to the Coharie formation found east of us. This sand layer varies in depth from 16′ to 1′. Underneath this sand layer is a layer of sand clay which varies in thickness from 2′ to 3′ and in some instances to a depth of 20 or more feet. Underneath this sand-clay layer we find the gravel and sand stratum. These vary in depth from 1′ up to approximately 50′ which is about as deep as we have ever excavated any of these layers.

8. *Hedrick Pits, Lilesville.*—These pits lie immediately south of the Bonsall pits and are essentially similar. A letter from B. V. Hedrick, President of the Company, states the following:

The deepest we have found gravel here is 20′ of overburden, 60′ of gravel, making a total of 80′. This is not an average depth by any means, but this is the thickest gravel that we have ever dug. The gravel usually contains around 10 to 20% in clay and 20 to 25% in sand. We find some layers of very tough clay anywhere from a few inches thick to two or three feet thick. In some places we have a deposit of gravel where the gravel is on top of the ground without any overburden, and some places the overburden runs as much as 20 to 25′ thick.

According to Mr. Hedrick, some fossils were found about 15 years ago, but no records are available on the matter.

CORRELATION

Because of the absence of fossils it is difficult to determine the correlation of these high-level gravels. The coarse gravelly nature of the sediment and the irregular pre-"Lafayette" surface seem to suggest a fluviatile origin. Therefore the present writer regards these high-level gravels as partly equivalent to the Bryn Mawr gravel of eastern Pennsylvania and the Beacon Hill formation of New Jersey, which are usually regarded as of late Pliocene age. They may also include equivalents of the Brandywine gravel of Maryland, which Cooke (personal communication) now regards as an alluvial form of Pliocene age.

EARLY PLEISTOCENE

On the Atlantic Seaboard, south of the terminal moraine, the Pleistocene formations are composed of gravel, sand, silt, and clay, ranging in thickness from a few feet to a score or more. They lie unconformably on the unconsolidated sediments of the Coastal Plain. At some places these form terraces, with what is considered to be a low wave cut bluff, or beach ridge at the landward margin. Cooke has described a series of these terrace remnants ranging from 25 to 270 feet above sea level. The higher ones are largely discontinuous and patchy; the lower ones are fairly continuous broad stretches of flat country, 30 to 50 miles wide. In North Carolina the various terraces are regarded by Cooke as coextensive with formations of the same name.

Three main theories have been advanced to explain the origin of these terraces: one that they are largely of marine origin, another that they are predominantly of fluvial derivation, and thirdly that there is the combination of the two.

Johnson (1907), Stephenson (1912) and Cooke (1930, 1931, 1932 and 1936) believed that the North Carolina terraces are largely of marine origin, while Wentworth (1930) and Campbell (1931) thought that the higher terraces (or their equivalents in Virginia) were of fluvial origin.

The marine origin of the lowest terrace formation (Pamlico) is easily demonstrated because of the abundant marine fossils. The origin of the higher "terraces" can only be determined by physiographic studies since no fossils, other than a few plant remains, are known from North Carolina above the 25-foot level.

Antevs (1929) and Cooke (1930, etc.) have applied the principle of glacial control to these terraces. Inasmuch as the glacial ages were times of low sea level and the interglacial ages times of high sea level, it was suggested that the terraces date from various interglacial ages. The following table shows a revised correlation proposed by Cooke and is published by his permission.

Recently Flint (1940, 1942) has criticized some of the interpretations of Cooke and has presented evidence to show that some of the higher "terraces" are of fluvial

TABLE 7

PLEISTOCENE TERRACE (after Cooke)

Altitude of Shore line (feet)	Name of Terrace	Age
?	———	Nebraskan glacial
215 170	Coharie Sunderland	Aftonian interglacial
?	———	Kansan glacial
140 100 70 42	Okefenokee Wicomico Penholoway Talbot	Yarmouth interglacial
?	(Horry clay)	Illinoian glacial
25	Pamlico	Sangamon interglacial
?	———	Wisconsin glacial

origin, and that only two definite shore lines are demonstrated. He has not correlated these exactly with any of the terrace names but has used Wentworth's name Suffolk scarp for a shore line approximately equivalent to the inner edge of the Pamlico. Its toe is between 20 and 30 feet elevation. The higher shore line (Surry scarp of Wentworth) has an elevation at its toe of 90 feet. Flint regards this feature as older than the Suffolk scarp. The marine origin of the lower scarp is confirmed by the presence of marine fossils (Pamlico formation); however, the evidence for the marine origin of the Surry scarp is largely based upon physiographic data, although marine diatoms from McBeth, S. C., were regarded by Flint as indicating a Pleistocene age.

In the present survey relatively little attention has been paid to the Pleistocene deposits older than the Pamlico, and no attempt has been made to redefine the terraces. Nor will any attempt be made to determine the origin of the formations, although it might be said that the present writer, being primarily a paleontologist, likes actually to see marine fossils before he is entirely convinced of the marine origin of a formation.

A few typical older Pleistocene exposures are recorded below and are referred to the formation as defined by Cooke. For further discussion of these Pleistocene deposits the reader is referred to Stephenson's discussion in Volume 3. It is possible that further work may show only a single formation of Early Pleistocene age, the various formations listed below being reduced to the rank of members.

BRANDYWINE FORMATION

The formation was named by W. B. Clark in 1915 from the town of Brandywine in Prince Georges County, Maryland. Cooke (1930, 1931) correlated the Brandy-

wine formation with part of the "Lafayette" and regarded it as of early Pleistocene age. He believed that it was of marine origin with a shore line at an elevation 270 feet above sea level. Cooke now (personal communication) regards the Brandywine gravels of Maryland as an alluvial fan of Pliocene age (see page 36). The gravels in North Carolina formerly assigned to the Brandywine formation (Richards, 1943) are here referred to the "Higher Gravels" of Pliocene age.

COHARIE FORMATION

NAME

Named by Stephenson (Vol. 3) from Great Coharie Creek, a tributary of Black River in Sampson County, North Carolina.

The extent of this formation is shown on plate 13 of Stephenson (1912). The "shore line" or contact between the Coharie and the Brandywine can be seen near Smithfield. A contact with the Black Creek formation is clearly shown along highway 701 at Stone Creek in Johnston County.

DETAILED SECTIONS

Johnston County

1. *Stone Creek.*—The Coharie is well exposed along Highway 701 between Clinton and Smithfield at Stone Creek bridge. It unconformably overlies the Black Creek.

Stephenson (Vol. 3: 275–276) records similar exposures in the same region.

Cumberland County

2. *Fayetteville.*—Pebbly sand of the Coharie formation overlies the Black Creek along Fort Bragg Highway between Fayetteville and Fort Bragg.

Nash County

3. *Nashville.*—The Coharie rests on the basement rocks in railroad cuts near Nashville (Stephenson, Vol. 3: 277).

Johnston County

4. *Kenly.*—"Good exposures showing the Coharie formation resting directly on the basement rocks may be seen in eastern Johnston County, especially in railway cuts in the vicinity of Kenly" (Stephenson, Vol. 3: 277).

SUNDERLAND FORMATION

NAME

Derived from the village of Sunderland in Calvert County, Maryland; named by Shattuck (1901).

EXTENT

As mapped by Stephenson, the Sunderland formation occupies part of Northampton, Halifax, Nash, Edgecombe, Wilson, Wayne, Johnston, Duplin, Sampson,

Cumberland, Bladen, Robeson, and Columbus counties. It forms a flat plane varying in width from between 5 and 10 miles near the Virginia line to more than 30 miles near the South Carolina border. According to Cooke (1931: 507) the Sunderland "shore line" can be traced across Kenly (N. C.) quadrangle at an altitude of about 170 feet. However, the existence of this "shore line" has been questioned by Flint.

DETAILED SECTIONS

Edgecombe County

1. *Rocky Mount.*—The Sunderland overlies the Miocene (Yorktown) in the Sewage Disposal Pits just east of Rocky Mount. A few pieces of fossil wood were collected.

A similar relationship was noted by Stephenson (Vol. 3: 278) in a road cut 5 miles south of Rocky Mount.

WICOMICO FORMATION

NAME

Named by Shattuck (1901) from the Wicomico River in Charles and St. Mary's Counties, Maryland.

EXTENT

As shown on Stephenson's map, the deposits of the Wicomico formation occupy a belt lying to the northeast of the Sunderland, although part of the area mapped as Wicomico may belong to the Penholoway which was not differentiated at the time of Stephenson's report. The shore line, according to Cooke, is at an elevation of 100 feet. The marine origin of the formation, however, is questioned by Flint.

DETAILED SECTIONS

Northampton County

1. *Weldon.*—About 1¼ miles east of Weldon and a short distance east of the bridge over Roanoke River, the following section was reported (Stephenson, Vol. 3: 281):

	Feet
Pleistocene (Wicomico formation)	
Mostly concealed by vegetation, but in part yellow, more or less sandy clay	35
Yellow, pebbly, argillaceous sand	2
Small lens of yellow, sticky clay, containing well-preserved prints of leaves	½
Yellow, very coarse, pebbly, arkosic sand. The pebbles are for the most part only partially waterworn	2½

The following plants were identified by Berry (1909, 1926): *Cercis canadensis* L.; *Liriodendron tulipefera* L.; *Quercus predigitata* Berry; *Quercus* sp.

Halifax County

2. *Palmyra Bluff.*—The upper layer of this bluff on the Roanoke probably belongs to the Wicomico formation (see page 24).

Pitt County

3. *Conetoe Swamp.*—Excavations for drainage canals in this swamp in most places passed through only the Wicomico formation. In a few places, underlying Miocene and Cretaceous deposits were encountered (see page 26).

Edgecombe County

4. *Old Sparta.*—The following section was noted 1 mile southwest of Old Sparta (after Stephenson, Vol. 3: 282):

	Feet
Pleistocene (Wicomico formation)	
Yellow, sandy clay, grading down into very coarse, yellow argillaceous sand, cross-bedded in places at base	8
Unconformity	
Miocene	
Yellow and red, fine sand, and drab clay	17

PENHOLOWAY FORMATION

NAME

Named by Cooke (1925) from Penholoway Creek and Bay in Brantley County, Georgia.

EXTENT

This formation has not been differentiated from the Wicomico in North Carolina. According to Cooke (1931a: 509), "the seashore of the Penholoway terrace forms the southern part of the Kinston (North Carolina) quadrangle." Cooke (1936: 333) indicates no erosion interval either before or after the Penholoway.

It is from this terrace deposit in Berkeley County, S. C., that the marine fossils were recently obtained (Richards, 1943b). The fauna may be analyzed as follows:

Pliocene and Pleistocene	7 species
Pliocene and not Pleistocene	6 species
Pleistocene and not Pliocene	4 species

Three possible interpretations of the fauna were suggested: (1) a mechanical mixture of Pliocene and Pleistocene deposits; (2) a late Pliocene age; (3) an early Pleistocene age. Interpretations (2) and (3) seemed more probable; if (1) or (3) were true, it would mark the highest level where marine Pleistocene fossils had been reported anywhere along the Atlantic Coastal Plain.

TALBOT FORMATION

NAME

Named by Shattuck from Talbot County, Maryland (1901). This formation included not only the Talbot in its present restricted sense, but all Pleistocene deposits between about the 45 feet contour and the present sea. A lower terrace deposit—the Pamlico with shore line at 25 feet—was later recognized.

Stephenson (1912), in describing the Pleistocene of North Carolina, divided the Talbot terrace into two parts, an upper "Chowan" named from the Chowan River, with the same shore line as the Talbot, and a lower Pamlico. Cooke (1931a: 510) has retained the older term Talbot in place of the "Chowan."

EXTENT

This terrace deposit occupies a belt southeast of the Penholoway. The northwestern extent of the formation is generally limited by a scarp or "shore line"; however, the Talbot terrace and formation extend up numerous rivers cutting into the older deposits. The eastern boundary of the Talbot is indicated by a fairly well marked scarp separating it from the Pamlico.

DETAILED SECTIONS

Bertie County

1. *Chowan River.*—Sandy deposits of this formation overlie the Miocene at numerous places along the Chowan River and immediately west. This is the region of the typical "Chowan" formation (= Talbot).

Beaufort County

2. *Washington.*—Similar outcrops of sand can be seen along highway 264 just east of Washington.

Pitt County

3. *Dupree Landing.*—Stephenson (Vol. 3: 286) recorded the following section on the Tar River, 14 miles above Greenville.

Feet

Pleistocene (Chowan formation)
Sandy loam and argillaceous sand with pebbles along base 6
Lignite bed consisting of a mass of logs, twigs, leaves, acorns, etc., in the form of lignite. Fills a depression in the underlying Patuxent formation 0–3
Unconformity
Lower Cretaceous (Patuxent formation)
Compact sands and clays 6

Berry recorded cones of *Pinus rigida* (Pine) and *Quercus phellos* (oak) from the lignite bed.

Wayne County

4. *Near Seven Springs.*—Stephenson and Berry have obtained fossil plants from a section on the right bank of the Neuse River 79⅔ miles above New Bern and about 4½ miles above Seven Springs.

Brunswick County

5. *Old Brunswick.*—Stephenson (Vol. 3: 284) recorded a section on the Cape Fear River ¼ mile north of the ruins of Old Brunswick and at another place 2 miles north of Southport. The bluffs here range between 15 and 35 feet in height. Stephenson referred these to the "Chowan" formation (= Talbot), but the present author recently suggested that they might well be part of the Pamlico (Richards, 1936a: 1636). Marine fossils were noted at the Old Brunswick locality.

HORRY CLAY

NAME

Named by Cooke (1937) from a section of the Intracoastal Waterway canal, 2½ miles northwest of Myrtle Beach, Horry County, South Carolina.

EXTENT

Known only from the type locality and from a site along the Neuse River, 10 miles below New Bern, Craven County, N. C. In both places a thin clay deposit lies beneath the Pamlico formation. Near Myrtle Beach the clay contains cypress knees, other plant remains, and numerous diatoms. According to Cooke (1937), "the presence of rooted tree stumps beneath a thick marine deposit that evidently accumulated in quiet water gives conclusive evidence that the sea stood lower on the land when they grew than in the immediately succeeding epoch."

In the clay along the Neuse River cypress stumps and diatoms are also present, and because of the similarity in stratigraphic position, it is here suggested that this clay may be correlated with the Horry clay of South Carolina and was deposited during a low stand of the sea (glacial stage) prior to the deposition of the Pamlico. For section of this locality see page 41 under Pamlico formation.

UNDERLYING FORMATIONS

At both localities the Horry clay is apparently underlain by the marine Pliocene (Croatan in North Carolina and Waccamaw in South Carolina). However, the actual contact was not observed.

OVERLYING FORMATIONS

At both places the Horry clay is overlain by fossiliferous Pamlico deposits.

THICKNESS

About 4 feet at the Neuse River locality and 3 feet near Myrtle Beach, South Carolina.

LATE PLEISTOCENE

PAMLICO FORMATION

NAME

In the 1912 (Vol. 3) report Stephenson divided the Talbot terrace and formation of Shattuck into two parts, an older one which he called "Chowan" (= Talbot) with shore line at about 42 feet and a lower one with shore line at about 25 feet which he designated as Pamlico from Pamlico Sound, North Carolina.

EXTENT

This lies east of the Talbot formation and covers that portion of North Carolina under the 25 (or 28) foot contour. It is extensively developed in extreme eastern North Carolina and is found in the following counties:

Currituck, Camden, Pasquotank, Perquimans, Tyrrell, Dare, Hyde, Beaufort, Pamlico, Craven, Carteret, Onslow, New Hanover, and Brunswick. South of Carteret County the Pamlico formation is less developed and usually occurs as a thin mantle on top of older deposits. However, near Cape Fear, in New Hanover County, it is better developed and locally occurs in the form of coquina.

The Pamlico terrace and formation has been traced with a remarkably horizontal shore line (25 or 28 feet) from New Jersey to Florida (Richards, 1936a), and also along the Gulf of Mexico (Richards, 1938, 1939). It is thought to date from the last major interglacial stage (Sangamon).

UNDERLYING FORMATIONS

In the Dismal Swamp area the Pamlico is underlain by Pliocene (?) or Miocene. In Hyde, Dare, Tyrrell, and Pamlico Counties it is apparently underlain below the surface by Pliocene (Waccamaw or Croatan), although there may be some older Pleistocene in between. Along the Neuse River and near Beaufort, it is underlain unconformably by the Croatan, except at one point 10 miles below New Bern where there is a thin intermediate layer of Horry clay (see page 39). In the Wilmington-Cape Fear region the underlying formation has not been determined.

OVERLYING FORMATIONS

Recent dune sands and peat overlie the Pamlico at various places. Post-Pamlico peats are abundant in Hyde County (see Richards, 1936a, esp. pp. 1632, 1645, and plate 4), the Dismal Swamp (Lewis and Cocke, 1929), and the "Open Lands" of Carteret County (Dachnowsky-Stokes and Wells, 1929). In the Cape Fear area some Post-Pamlico peats and sands are being studied by Dr. B. W. Wells of North Carolina State College.

THICKNESS

Usually between 15 and 20 feet in thickness. Along the north shore of the Neuse River, 16 miles below New Bern (Bennett Plantation), a thickness of about 28 feet was noted, apparently the entire bluff being of Pamlico formation. Similarly at a cut of the Intracoastal Waterway Canal near Carolina Beach, a thickness of at least 30 feet was noted. Here the coquina extended up to 22 feet above sea level.

LITHOLOGY

Sandy loams, sand, and clay. Locally, near Cape Fear, there is a coquina bed that outcrops along the beach and which is encountered in excavations and quarries near Carolina Beach. It is locally used for road building. The extensive shell deposits struck in dredging the Intercoastal Waterway canal in Hyde and Pamlico Counties, and elsewhere are occasionally used for highways, but because of the sandy matrix are rarely used for agricultural purposes.

DETAILED SECTIONS
Camden County

1. *South Mills.*—Excavations for the Dismal Swamp Canal passed through the Pamlico formation and numerous fossils were obtained from the spoil banks. At the present time the best material can be seen about 5 miles north of South Mills and 2 miles south of the Virginia line. In many places the shells have been hauled away for road building or for other purposes. The presence of many shells of *Arca subsinuata* Conrad may indicate that the Pliocene is not far below and that these shells may have been reworked from that formation.

Perquimans County

2. *Nicanor.*—Canals along a road leading from Nicanor toward the Pasquotank County line, near the western border of the Dismal Swamp, show an exposure of shells about six feet below the surface of the land. The great majority of the shells are of the oyster (*Ostrea virginica*), and many of the individuals are of unusually large size. In addition to the oyster, numerous other species are present such as: *Arca transversa, Venus mercenaria, Mya arenaria, Macoma balthica, Nassa obsoleta, N. trivittata, Urosalpinx cinerea, Eupleura caudata.* The presence of the northern species *Mya arenaria* and *Macoma balthica* is of considerable interest since these species are rare or absent from the Recent seas of North Carolina, and have only rarely been seen by the writer from the Pleistocene of the state. The assemblage suggest brackish water such as in the vicinity of an oyster reef and probably represents a brackish phase of the same age as that from the eastern part of the Dismal Swamp (South Mills).

3. *Bear Swamp.*—The late Mr. L. W. Anderson, County Agent of Perquimans County, informed the writer that oyster shells were struck some years ago in excavations for canals in Bear Swamp between Hertford and Tyner.

Pasquotank County

4. *Elizabeth City.*—Wells in this vicinity show that the Pamlico extends from near the surface to a depth of about 30 feet (see Henbest, Lohman, and Mansfield, 1939).

Hyde County

5. *Intracoastal Canal.*—Shells are abundant from the spoil banks of the land cut of the Intracoastal Waterway canal between Pungo and Alligator Rivers. In most places the shell layer is overlaid by peat, but nowhere are the shells actually seen above the surface. About a hundred species were obtained from the spoil banks. For further details see Richards (1936a: 1632) (fig. 51).

6. *Swan Quarter.*—Similar Pleistocene shells have been obtained from shallow ditches near Swan Quarter. They have also been noted from shallow wells here and elsewhere in Hyde County.

7. *Lake Matamuskeet.*—Marine fossils were obtained in 1941 from excavations in connection with the road across the Lake from Fairfield to New Holland. All specimens examined were of Pleistocene age with the exception of one individual of *Ostrea compressirostra* Say which might have been reworked from the underlying Pliocene. The marine Pleistocene is covered with at least 6 feet of peat.

8. *Lake Landing.*—A well drilled near here showed the Pamlico formation extending from the surface down to — 80 feet (Stephenson, Vol. 3: 252).

9. *Ocracoke.*—Numerous Pleistocene and Pliocene shells have been washed onto the beach on Ocracoke Island (see Richards, 1936a: 1637).

10. *Manteo.*—Some Pleistocene fossils were obtained from a fill in the town of Manteo, which had been pumped by hydraulic dredging from beneath the bottom of the harbor. Some of the shells obtained were of Pleistocene age. Among these were several individuals of *Neptunea stonei* (Pilsbry) (fig. 76b), a species of northern affinities that had not hitherto been reported from the Pleistocene or Recent of North Carolina. The species has been found in New Jersey, Massachusetts and Long Island in deposits correlated with the Wisconsin glacial stage. It is generally believed that the glacial shore line was some miles east of the present beach; therefore, the presence of these fossils at Manteo is somewhat surprising. There may have been an inlet in the Manteo region similar to one recently suggested as having occurred during Wisconsin time in the vicinity of Cape May, New Jersey. The other species obtained from the Manteo fill included *Crepidula fornicata* L. and *Cruciculum costatum* Linné. These are probably post-Pamlico (Wisconsin) in age.

11. *Stumpy Point.*—A few shells were obtained from shallow pits along the Englehard-Stumpy Point road, about 2 miles from Stumpy Point.

Tyrrell County

12. *Kilkenny.*—A few shells were found in borrow pits on the Gum Neck Road, 2 miles north of Kilkenny.

Pamlico County

13. *Bayboro.*—According to Dr. H. N. Coryell of Columbia University, fossil shells were obtained "north and a little east of Bayboro."

14. *Hobucken.*—Shells were found on the spoils banks of the Intracoastal Canal which passes through a small land cut about 1 mile west of Hobucken.

15. *Cash Corner.*—This entire region appears to be underlaid by the marine Pamlico formation. Several small canals near Cash Corner and Alliance cut into the shell beds and material was obtained from the spoil banks.

16. *Benners Plantation.*—Croom (1835) and Conrad (1835) described some marine fossils from marl pits on the Lucas Benners Plantation on the left bank of the Neuse River, 16 miles below New Bern. In addition to the invertebrate fossils, a number of vertebrate fossils were reported from these pits (Hay, 1923).

This property is now owned by Dr. Don Lee of Arapahoe, North Carolina. It is on the east side of Bairds Creek.

17. *Bennett Plantation.*—Bluff along the Neuse River on the opposite side of Bairds Creek from the previous locality. Here the bluff is 30 feet high and the following section was noted:

	Feet	
Sand	12	
Clay with shells	6	} Pamlico
Sandy clay, shells	12	

This is now the property of Dr. D. A. Dees (fig. 50).

Craven County

18. *Ten Miles Below New Bern.*—The following section was recorded on the right (southwest) bank of the Neuse River near the settlement of Croatan (Mansfield, 1928; Richards, 1936):

	Feet
Loamy soil	2
Laminated, gray to reddish medium-grained sand	4
Compact, laminated gray clay with thin partings of sand; impressions of shells	4
Gray clayey sand	4
Very fossiliferous grayish sand; especially *Mulinia lateralis*; *Rangia cuneata* in lower 1 foot	8
Unconformity	
Truncated cypress stumps, 6 to 8 feet in diameter, embedded in dark carbonaceous clay	4
	26

The clay containing the cypress stumps is referred to the Horry formation, and the overlying material to the Pamlico (fig. 46, 47).

At the time of Mansfield's visit the land was owned by W. B. Flanner (U.S.G.S. 10896); at the present time it is part of Croatan National Forest.

19. *Eleven Miles Below New Bern.*—Mansfield (1928: 134) recorded the following section:

	Feet
Sandy soil	2
Pleistocene	
Laminated reddish sand and clay, with a water seepage at the base	6
Gray clayey sand	8
Very fossiliferous fine-grained sand	4
Unconformity	
Pliocene (Croatan sand)	
Concretionary, ferriginous coarse sand and gravel, carrying corals and mollusks	0–2

Erosion has destroyed much of this bluff and at the times of the present writer's visits (1934, 1940, 1941) the Pliocene was not exposed.

20. *Croatan.*—Pamlico shells were obtained from shallow marl pits on the left prong of Bryces Creek, 1 mile west of Croatan railroad station.

Mansfield recorded several localities in this vicinity; in some the Pliocene (Croatan) was mixed with the Pleistocene.

21. *Havelock.*—The upper 40 feet of the oil test at Lake Ellis 5 miles west of Havelock passed through the Pamlico (Mansfield, 1925).

Carteret County

22. *Beaufort.*—Canals for the Open Land Project about 10 miles northwest of Beaufort and 6 miles from North River yielded Pamlico fossils (Mansfield, 1928; Richards, 1936).

23. *Core Creek Canal.*—Excavations for the Intracoastal Waterway canal near Core Creek Bridge have revealed numerous fossils. Immediately southwest of the bridge, the fossils are all Pleistocene and referable to the Pamlico formation. One mile northeast of the bridge the spoil banks contain a mixture of Pleistocene (Pamlico) and Pliocene (Croatan) fossils.

Onslow County

24. *Stump Sound.*—A mixture of Pamlico and Croatan fossils was obtained from the spoil banks of the Intracoastal Waterway canal at Tar Landing, on Stump Sound, southeast of Folkstone. The majority of the fossils were Pleistocene.

New Hanover County

25. *Gander Point.*—In a gravel pit at Gander Point, between Wilmington and Carolina Beach, shell fragments and coquina are abundant. This locality is near the bridge over the Intracoastal Waterway canal.

26. *Near Carolina Beach.*—At the crossing on the Intracoastal Waterway canal of highway 421, Pleistocene coquina is exposed up to an elevation of 22 feet above tide. Nearby, this coquina is overlain by a layer of black sandstone and peat.

In the construction of this canal, the bones of a mastodon were uncovered.

27. *Old Fort Fisher.*—Considerable coquina is exposed between tides on the beach at Old Fort Fisher. This coquina is very similar to that of the Anastasia formation of Florida which is probably contemporaneous with the Pamlico (fig. 48).

Brunswick County

28. *Old Brunswick.*—Stephenson (Vol. 3: 1636) recorded a bluff along the Cape Fear River ¼ mile north of the ruins of Old Brunswick which he referred to the "Chowan" formation (= Talbot). The present writer regards it as belonging to the Pamlico. Fossil shells were present.

29. *Southport.*—Stephenson recorded fossils from the Pleistocene section of the well at the quarantine station near Southport.

30. *Fort Caswell.*—Stephenson also reported fossils from the Pleistocene section of the well at Fort Caswell, Oak Island, in Cape Fear River.

31. *Winnabow.*—Some years ago, Mr. G. C. Arp, of Winnabow, obtained some mastodon bones on top of some shell marl near his farm. The bones are now in the North Carolina State Museum.

32. *Holden's Ferry.*—Many shells of *Arca ponderosa* were collected from the spoil banks of the Intracoastal Canal at Holden's Ferry 7 miles south of Supply.

MISCELLANEOUS PLEISTOCENE VERTEBRATE LOCALITIES

The following localities have not been correlated with any specific Pleistocene formation. In most cases quoted from previous literature the information is too incomplete. Even in more recent collectings, it is not always easy to correlate a locality with one of the several formations. For further data see Hay (1923).

Halifax County

1. C. Cobb (1923: 31, 32) recorded *Equus* (horse) and a whale from somewhere in this county.

Pitt County

2. *Greenville.*—*Equus complicatus* was obtained near Greenville by Emmons.

Carteret County

3. *Core Creek Canal.*—Mammoth bones and Mastodon teeth were obtained from dredgings from the Intracoastal Waterway canal. The invertebrate fossils associated with these bones have been correlated with the Pamlico formation.

4. *Harlowe.*—Bones of a cetacean and Mastodon were obtained here.

Edgecombe County

5. *Tarboro.*—The United States National Museum has in its collection the bones of *Mastodon americanus* from Tarboro.

Nash County

6. *Rocky Mount.*—*M. americanus*—after Emmons.

Washington County

7. *Equus leidyi*—after Emmons.

Bladen County

8. *Elizabethtown.*—*E. leidyi*—after Emmons. According to Hay, this is probably in the Sunderland Formation

Craven County

9. *Fort Barnwell.*—A muck deposit on top of Miocene marl on the property of Z. B. Broadway contained teeth of the Tapir (*Tapirus haysii*), Mastodon and a cetacean (see page 26).

Jones County

10. *Maysville.*—Horse and Mastodon (North Carolina State Museum).

11. *Bender Farm.*—Horse and Manatee (North Carolina State Museum).

Pamlico County

12. *Minnesott Beach.*—The North Carolina State Museum has the following: *Carcharodon megalodon; C.* cf. *acutidens; Osyrhina* sp.

13. *Benners Plantation.*—Several vertebrate fossils have been recorded from this locality including *Mammut americanum, Elephas, Equus* and several cetaceans. For discussion of fauna see Hay (1923: 358–359). The marine shells associated with these fossils have been referred to the Pamlico formation (see page 41).

Onslow County

14. *Jacksonville.*—*Mastodon americanum* (N. C. State Museum).

Duplin County

15. Mastodon bones have been found on the William Hatcher Farm (N. C. State Museum).

Wayne County

16. *Mastodon americanum* (N. C. State Museum).

Pender County

17. *M. americanum* (N. C. State Museum).

New Hanover County

18. *Near Carolina Beach.*—Mastodon bones were dug from the Intracoastal Waterway canal (see page 42).

19. *Ross and Lord's Quarry.*—Mammoth (N. C. State Museum).

20. *Nine Miles South of Wilmington.*—*Elephas columbi* (N. C. State Museum).

21. *Wrightsville Beach.*—*Equus* (N. C. State Museum).

22. *Winnabow (Brunswick County).*—Mastodon.

Dare County

23. *Kitty Hawk.*—Bones of the walrus (*Odobenus rosmarus*) from the beach at Kitty Hawk are in the museum at Yale University. The species is thought to have lived in North Carolina waters during one of the glacial stages.

CORRELATION

With the exception of *Neptunea stonei,* which is probably post-Pamlico, all the invertebrate species reported from the Pamlico formation of North Carolina are still living today, although some of them live a little farther south. Thus, the fauna may be taken as indicating a slightly warmer climate than that of today. Similarly, the fauna of the Pamlico elsewhere along the southern

TABLE 8

CORRELATION OF LATE PLEISTOCENE OF ATLANTIC COASTAL PLAIN (AFTER RICHARDS, 1936a)

	NEW JERSEY Salisbury & Knopf, 1917	MARYLAND Shattuck, 1906	VIRGINIA Clark & Miller, 1912 Wentworth, 1930	N. C. Stephenson, 1912	S. C. Cooke, 1936	GEORGIA Cooke, 1925 Veatch & Stephenson, 1911	FLORIDA Matson, 1913; Leverett, 1931 Cooke, Mossom, 1929
McGee							Fort Thompson
			Princess Anne				Key Largo
			Dismal Swamp	Pamlico	Pamlico	Pamlico / Satilla	Miami
Columbia*	Cape May	Talbot	Talbot				Pensacola
			Chowan	Chowan	Talbot	Talbot	Anastasia

* Includes all Pleistocene Terraces.

Atlantic Coastal Plain from New Jersey to Florida indicates a slightly warmer climate than that prevailing in the same latitude today. For this reason, as well as for stratigraphic reasons detailed at length elsewhere (Richards, 1936a), the Pamlico formation has been dated from the last major interglacial stage of the Pleistocene.

Marine fossils have been found at and below the 28-foot level all the way from New Jersey to Florida and along the Gulf of Mexico to at least the Texas-Mexico border. It is believed that the 28-foot contour represents the interglacial shore line at the time that there was more water in the sea because of the melted glaciers. Table 8, adapted from an earlier work of the present writer, shows the various formation and terrace names that have been used for these late Pleistocene terrace deposits. South of New Jersey, at least through Georgia, the Pamlico terrace and formation seem to be continuous and therefore that name seems to be acceptable.

According to the glacial control hypothesis, the shore line during glacial times was probably some miles east of that of the present time. The presence of the Arctic *Neptunea stonei* at Manteo suggests that there may have been an embayment or inlet in that region that allowed the cold waters to have extended inland to the present site of Manteo.

FOSSILS FROM THE PAMLICO FORMATION

(See also list by Richards (1936))

Crustacea

Balanus crenatus Bruguière 29

Plececypoda

Nucula proxima Say 5, 23, 18, 29
Leda acuta Conrad 18, 23, 29
Yoldia limatula Say 5, 11, 18, 29
Arca campechiensis Say 5, 13, 15, 18, 22, 23
Arca ponderosa Say 5, 11, 14, 15, 18, 22, 23, 25, 27
Arca incongrua Say 5, 22
Arca transversa Say 2, 5, 11, 13, 14, 15, 18, 22, 23
Ostrea virginica Gmelin 2, 5, 13, 14, 15, 18, 22, 23, 25, 27
Pecten gibbus irradians Lamarck 5, 12 14, 22
Anomia simplex Orbigny 5, 18, 22
Pandora trilineata Say 18
Crasinella lunulata Conrad 5, 22
Venericardia tridentata Say 22, 5, 23, 27, 28
Phacoides crenella Dall 5, 14, 15, 23
Divarcella quadrisulcata Orbigny 5, 13, 14, 15, 22 (?)
Cardium robustum Solander 5, 13, 19, 23
Dosinia elegans Conrad 5
Macrocallista nimbosa Solander 5, 23
Callocardia morrhuana Linsley 5
Chione cancellata Linnaeus 5, 14, 15, 22, 23
Chione latilirata Conrad 5

Venus merceneria Linnaeus 2, 5, 11, 12, 14, 15, 18, 22, 23, 27
Venus campechiensis Gmelin 5, 15, 23
Tellina sayi Deshayes 18, 22
Tellidora cristata Recluz 5
Abra aequalis Say 18
Donax variabilis Say 5, 22, 23, 27
Tagelus gibbus Spengler 5, 11
Ensis directis Conrad 5, 13, 15, 18, 22, 23, 29
Mactra solidissima Dillwyn 5
Mulinia lateralis Say 5, 11, 13, 14, 15, 18, 22, 23, 27, 29
Corbula contracta Say 5, 22, 23, 29
Pholas costata Linnaeus 5, 13, 23

Gastropoda

Scalaria lincata Say 5, 18, 23
Pryamidella crenulata Holmes 5, 13, 14
Polinices heros Say 5
Sinum perspectivum Say 5, 18, 22
Crepidula fornicata Linnaeus 5, 15, 18, 22, 29, 30
Crepidula plana Say 5, 18, 22, 29
Crepidula convexa Say 5, 11, 14, 23
Eupleura caudata Say 2, 5, 18, 23
Urosalpinx cinera Say 2, 5
Columbella avara Say 5, 18, 23
Columbella lunata Say 5, 18, 23, 29
Columbella obesa Adams 18
Nasa obsoleta Say 2, 5, 11, 18, 23
Nasa trivittata Say 2, 5, 11, 13, 14, 15, 18, 22, 23, 29
Nasa vibex Say 5
Nasa acuta Say 5, 14, 15, 18, 23
Fulgur canaliculata Linnaeus 5, 13, 15, 18, 22, 23
Fulgur carica Gmelin 5, 11, 18, 23, 27
Fulgur perversa Linnaeus 5
Oliva sayana Ravenel 5, 12, 14, 15, 22, 23
Olivella mutica Say 5, 13, 14, 22, 23
Terebra dislocata Say 5, 14, 15, 18, 22, 23
Terebra concava Say 5, 15, 18
Acteocina canaliculata Say 5, 14, 15, 18, 23, 29

RECENT FORMATIONS

These include the deposits now being formed as well as those laid down since the end of the Wisconsin glacial period. In North Carolina this category includes the fluviatile and alluvial sands, silts, clays, and gravels now being formed by rivers and lakes; also the dune sands and other deposits of the barrier beaches of the "Banks"; and finally such swamp and other deposits that have accumulated on top of other sediments in relatively recent times.

Under this latter heading may be mentioned the extensive peat deposits of the Dismal Swamp and similar morasses of Tyrrell, Dare, Hyde, Pamlico, and Carteret Counties, and elsewhere. Studies on the post-Pleistocene history of these swamps have been made by Lewis and Cocke (1929), Dachnowski-Stokes and Wells (1929), and Richards (1936a: 1645).

The post-Pleistocene history of the peats and sands on top of the Pamlico coquina in the Cape Fear region is now being studied by Dr. B. W. Wells, of North Carolina State College. There is accumulating considerable evidence to show that there was a low stand of the sea following the deposition of the Pamlico formation. From this time forward there has been a rise to the present level. Whether this rise has been gradual or intermittent or cyclic still remains to be worked out.

Undeniable evidence of recent changes in shore line can be seen at various places along the North Carolina coast. Whether this be due to the sinking of the land, the rise of sea level, or merely to the erosive action of the waves and currents, is still a matter of some controversy. Cape Hatteras has receded more than a mile during the past century; on the other hand Cape Lookout has added almost as much land. From studies in New Jersey, the present author concluded that, while sea level was rising very slowly owing to the release of water from the melting glaciers, the main cause of the loss of land along that coast was the erosive action of the waves and currents (Richards, 1931, 1934). Probably the same thing is true regarding the North Carolina coast. Recent studies on the North Carolina beaches have been made by Hite (1924), Rude (1923), Hazlett (1938), and others (fig. 53).

THE "CAROLINA BAYS"

Considerable attention has been focused during recent years upon the so-called "Carolina Bays," marshy depressions, mainly oval in shape, that exist at numerous places along the coast of North and South Carolina; they are probably best developed in the region between Wilmington, N. C., and Myrtle Beach, S. C.

The attention of geologists was first called to these bays in 1884 when Dr. L. C. Glenn wrote a report on the oval bays near Darlington, S. C. No positive explanation of the origin of the bays was given at that time, although Glenn suggested that their origin may have been connected with the eastward retreat of the sea.

The bays were not brought to the attention of geologists again until 1932, when Melton and Schriever proposed a meteoric origin of these depressions; in other words they suggested that these oval basins and their rims were formed many years ago by an impact of a shower of meteors upon dry land. This theory attracted world-wide interest, and from 1932 to the present day at least 36 scientific papers have been written on these bays, as well as numerous popular articles.

Many geologists attacked the theory of meteoric origin, while others rallied to its support. For example, Cooke (1934) suggested that the elliptical sand ridges were in part bars and beaches built up in shallow lagoons during a higher stand of the sea, and in part crescent-shaped keys built up in shallow lakes. The parallelism of the oval depressions was explained by "a constancy in the direction of the wind while they were being shaped."

Later (1940) Cooke discarded the hypothesis of uniform wind and attempted to explain both form and orientation as due to "the tendency of rotary currents in liquids to assume the shape of an ellipse whose major axis points N 45° W in the Northern Hemisphere and N 45° E in the Southern Hemisphere."

Support for the meteoric hypothesis was contained in papers by Prouty, McCarthy, Straley, and others, in which the results of magnetometer surveys are discussed. According to Prouty (1935), "most of the elliptical bays and elliptical lake basins show a decided high magnetic area to the southeast of the bays." Prouty believed that this indicated buried meteoric bodies. Later work did not reveal conclusive evidence of buried meteoric material. McCarthy (1937) suggested that the crater-like depressions were formed by shock-waves of air accompanying a shower of large meteorites rather than by the meteorites themselves, and that the magnetic material underground represented condensations from the vapor formed when the meteorites volatilized.

Douglas Johnson (1934) criticized the meteoric theory and suggested that the bays in the Myrtle Beach area represented earlier freshwater lakes formed on a beach plain approximately at the present level of the bays, and that the oval rims represented accumulations of wind blown sand. This theory was expanded (Johnson, 1936), and it was further shown that many of the bays were associated with sinkholes or other solution phenomena. In 1937 Johnson further expanded his theory by showing that artesian springs, fed by shallow uprising artesian waters before stream incision in the Coastal Plain had lowered ground water, would have produced basins or craters similar to the Carolina Bays, partly by solution and partly by removal of finer sediment.

This theory was still further developed in 1940 and especially in a book published in 1942. In the latter work (1942: 154) Johnson summarizes his "artesian-solution-lacustrine-aeolian hypothesis," or more simply his "hypothesis of complex origin" as follows:

It supposes that artesian springs, rising through moving ground water and operating in part by solution, produced broad shallow basins occupied by lakes, about the margins of which beach ridges were formed by wave action and dune ridges by wind action.

In addition to discussing this hypothesis at great length in his book, Johnson also presents arguments against the meteoric hypothesis as well as the various other suggestions. He also gives an excellent summary of the entire problem.

The artesian part of the complex hypothesis has been criticized by McCampbell (1944, 1945) and Melton (1950), and many geologists still favor the meteoric hypothesis. Even the age of the bays is a matter of controversy. Estimates concerning their age vary from the Pleistocene to the late Cretaceous. So, in spite of the many papers written on the subject, the problem of the Carolina Bays must still be regarded as unsolved.

STRUCTURE OF NORTH CAROLINA COASTAL PLAIN

A few years ago it was generally believed that the basement rock of the Coastal Plain sloped rather gently and uniformly from the "Fall Line" toward the present ocean. Because of the scarcity of wells and the lack of geophysical data no detailed information was available.

With the increase of test drilling and the more frequent use of geophysical studies, geologists have been able to obtain considerably more information on the structure of the basement, and it is now realized that the slope is not as simple as was formerly believed. In 1935 a grant was awarded from the Penrose Bequest of the Geological Society of America to B. L. Miller and Maurice Ewing for a geophysical investigation across the emerged and submerged portions of the Atlantic Coastal Plain. Three seismic traverses were run: (1) "Barnegat Bay Section" from Plainsboro, N. J., to Silverton, N. J.; (2) "Cape May Section" from Bridgeport, N. J., to Avalon, N. J., and (3) the "Cape Henry Section" from Petersburg, Va., to Cape Henry, Va., and thence eastward across the Continental Shelf to the inner edge of the Continental Slope. The latter traverse has a direct bearing on problems of the North Carolina Coastal Plain. For a general discussion of methods see Ewing, Crary, and Rutherford (1937) and for a discussion of the geological interpretations of the geophysical data see Miller (1937).

Miller (1937) pointed out that the well records showed that the basement rock dipped at a rate of about 30 feet per mile in eastern Virginia, while in North Carolina the dip was only 10 feet per mile. The geophysical results showed a dip of about 30 feet per mile from the "Fall Line" near Petersburg, Va., to a point about 20 miles east of Cape Henry, Virginia. Beyond this point the slope increases, so that at a point approximately 60 miles at sea, off Cape Henry, the basement may be 12,000 feet deep.

The geophysical data suggested that there were two main zones in the sediments overlying the basement crystallines, particularly east of the present shore line. The upper zone was thought to include Cretaceous and younger rocks, while it was suggested that the lower zone might include some Jurassic and Triassic sediments, part of which might have been marine.[4]

---

[4] Marine Jurassic ostracods have recently been reported from the well at Cape Hatteras, North Carolina (Swain, 1950). Shales and sandstones, apparently non-marine, tentatively assigned to the Triassic, have been found in deep wells near Salisbury and Berlin, Maryland (Richards, 1945a, 1948a).

In comparing the Virginia and New Jersey traverses, it was later pointed out (Ewing, Woollard, and Vine, 1939: 285) that the seismic break in Virginia might well be the "M-Zone" of New Jersey which they interpret as the contact between the Magothy and Raritan formations. It was pointed out by the present author (Richards, 1945a: 951) that it might be better to interpret the "M-Zone" as the top of the Magothy rather than the Magothy-Raritan contact. There is much more definite break between the Magothy and the next overlying formation (Merchantville formation) in New Jersey than between the Magothy and the Raritan formations. Furthermore, there is commonly a zone at the base of the Merchantville formation which contains many ironstone crusts which might give the effect of a consolidated bed in geophysical tests. On the other hand, the Magothy-Raritan contact is very difficult to recognize lithologically or faunally in southern New Jersey and southward.

A few irregularities in the slope of the basement of rock of North Carolina have been known for many years. At least two outcrops of granite are known east of the "Fall Line." Both east and west of Fountain, Pitt County, the granite is about 400 feet deep, whereas in the town of Fountain there is a granite quarry (now abandoned) indicating a "hill" in the basement (see fig. 59). A similar "hill" surrounded by Coastal Plain sediments occurs near Smithfield (fig. 59, 60).

The drilling of the deep oil test at Cape Hatteras showed the basement rock to be considerably deeper (9,878 feet) than had been anticipated, and it became apparent that the slope increased markedly toward the east. It has recently been shown (Prouty, 1946; Berry, 1948) that the slope from the "Fall Line" to approximately Havelock, Craven County, averages about 14 feet per mile, while from New Bern to Cape Hatteras it is about 122 feet per mile. Prouty has tentatively referred the more gently sloping plane to the Schooley Peneplane and the deeper one to the Fall Zone peneplane (fig. 8).

A contour map of the Basement rock of the North Carolina Coastal Plain is shown in figure 10.

It was formerly suggested by the writer (Richards, 1945b) that another such hill existed near Havelock,

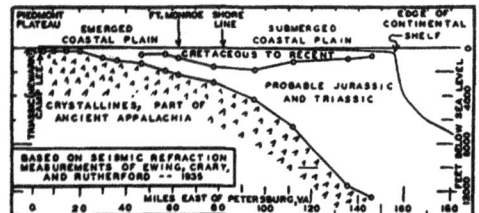

FIG. 7. Section across Coastal Plain from Petersburg, Va., through Fort Monroe, Va., to outer edge of Continental Shelf (after B. L. Miller, *Bull. G. S. A.* 48: 804).

Craven County. The well of the Great Lakes Drilling Company reached basement at 2,318 feet, while the more recent wells of the Carolina Petroleum Company near Merrimon, Carteret County, did not reach basement until a depth of about 4,000 feet. It was thus suggested that the basement rock might slope off rather smoothly, although increasing in dip, and that the Havelock well might represent the top of a hill. However, the data from the additional wells drilled by the Carolina Petroleum Company (especially the N. C. Pulpwood, Atlas Plywood, and Lindley) suggest a greatly increased slope from Havelock toward the east, and are thus more in line with the diagrams of Prouty and Berry.

The most significant structural feature of the rock floor of North Carolina Coastal Plain is the "Great Carolina Ridge" or the "Cape Fear Arch" which occurs in the vicinity of Cape Fear. In this region the basement rock extends to within about 1,100 feet of the

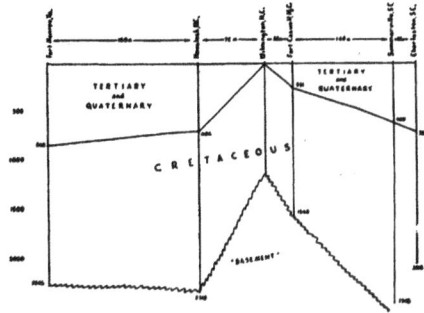

FIG. 8. Generalized cross section Durham, N. C., to Cape Hatteras, N. C. (after W. F. Prouty, *Bull. A. A. P. G.* 30: 1918).

surface whereas both north and south of the "ridge" it drops to a depth of 4,000 feet or more. The "ridge" may represent an irregularity in the surface of the basement rock caused by topography, or, more probably, it may represent a Cretaceous (?) folding of the basement. In any event, the Cretaceous and Tertiary formations "feather out" on the flanks of the uplift (see fig. 9).

There are indications of the existence of a basin between Cape Fear and Cape Henry. The southern limit is well marked by the "Great Carolina Ridge" and there are indications that there is also a rise in the basement in the vicinity of Cape Henry, Virginia, although the lack of deep wells in this area renders this interpretation highly tentative.[5]

As a result of magnetic studies, W. R. Johnson believed that this basin is limited on the west by a ridge,

[5] Basement was encountered in a well at Fort Monroe, Va., at a depth of 2,246 ft.

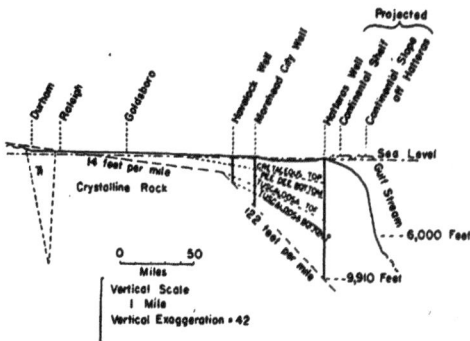

FIG. 9. Generalized cross section from Fort Monroe, Va., to Charleston, S. C., showing "Great Carolina Ridge" (after Richards, *Bull. A. A. P. G.* 29: 942).

extending roughly between Plymouth, Washington County, to Edendon, Chowan County, probably composed of low hills of gabbroic or similar rocks. It was also suggested that there is a "valley" lying between this ridge and the Fall Zone. These data indicate irregularities in the basement floor of North Carolina.[6]

In 1935 and 1936 McCarthy and Straley conducted geomagnetic investigations in eastern North Carolina. They observed a series of anomalies elongated subparallel with Appalachian tectonic trends. They thought it probable that the source of the anomalies plunged southwestward, and attributed them to either structural conditions or topography in the pre-Mesozoic basement.

The geophysical data on North Carolina were recently summarized by Straley and Richards (1949) in a paper presented before the International Geological Congress in London in 1948.

The Standard Oil Company of New Jersey conducted an extensive geophysical (seismic, gravity, magnetic) survey preparatory to drilling their test wells near Cape Hatteras. However, the information has not yet been released.

Additional off-shore seismic work, under the direction of Dr. Maurice Ewing, has been carried on during the past few years. Undoubtedly, considerable valuable information will be available when the full results of this work are published. In a preliminary abstract Ewing, Worzel, Steenland, and Press (1946: 1192) report that seismic refraction measurements were made from the coast line to the edge of the continental shelf along three lines of traverse, namely near Cape May, N. J., New York, N. Y., and Woods Hole, Mass.

[6] Johnson presented a paper before Section E of the American Association for the Advancement of Science in Richmond, Virginia, in December, 1938, and an abstract was published (Johnson, 1938). However, the full report never appeared in print. Some of the information contained in the paper was summarized by Straley and Richards (1949). Another summary by Straley will appear in a publication of the Department of Mines, Mining and Geology of Georgia.

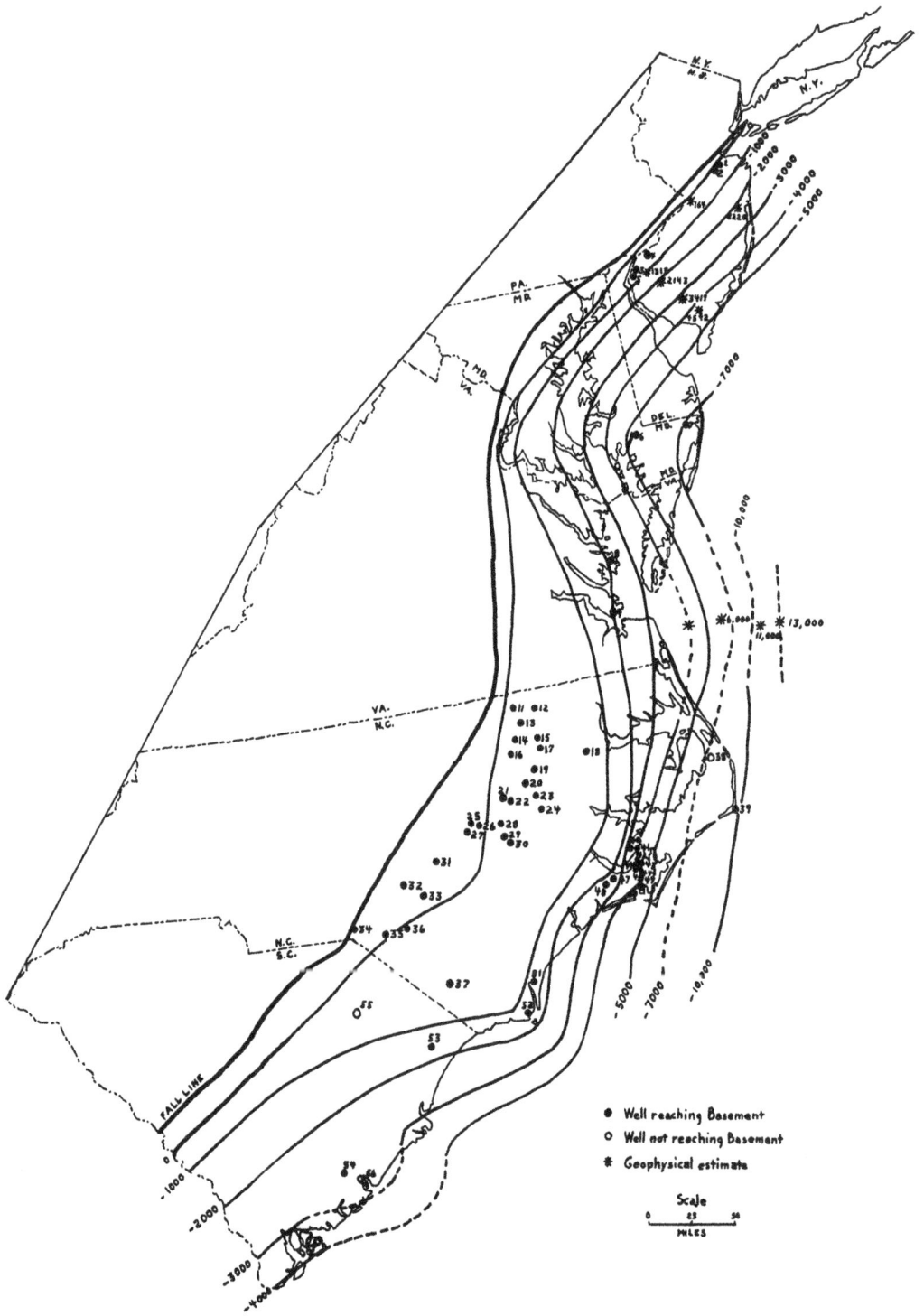

FIG. 10. Contour map of "Basement complex." New Jersey to South Carolina.

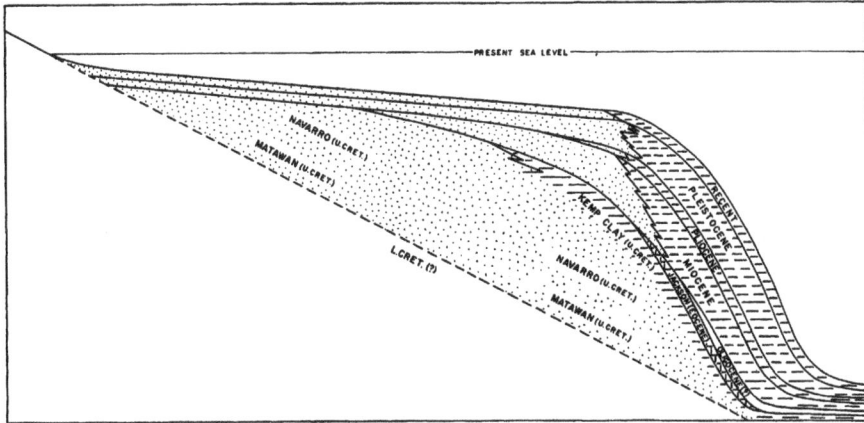

Fɪɢ. 11.　Idealized transverse section through Continental Shelf and Slope (after Stetson, *Papers in Physical Oceanography and Meteorology* 11 (2) : 25, 1949).

They report that the Cape May traverse agrees roughly with the Cape Henry traverse of 1935, and that the well at Cape Hatteras which reached basement at 9,878 feet confirms the seismic results.

Recently Stetson (1949) presented the results of studies on the Atlantic continental shelf between Georges Bank and Norfolk canyon and has summarized the information on the fossils obtained by dredgings from the bottom of the ocean. No deposits older than

Upper Cretaceous were reported by Stetson (p. 25, fig. 4).

ECONOMIC GEOLOGY

SAND

Sand is found rather generally throughout the Coastal Plain of North Carolina. Because of the wide distribution there are few extensive pits, limited local supplies usually being available. Probably the greatest

1. Old Bridge, Middlesex Co., N. J.	− 280
2. Spottswood, Middlesex Co., N. J.	− 280
3. Carneys Point, Salem Co., N. J.	− 268
4. Paulsboro, Gloucester Co., N. J.	− 271
5. Deepwater Point, Salem Co., N. J.	− 403
6. Salisbury, Wicomico Co., Md.	− 5472
7. Berlin, Worcester Co., Md.	− 7142
8. Matthews, Matthews Co., Va.	− 2305
9. Fort Monroe, Elizabeth City Co., Va.	− 2236
11. Weldon, Halifax Co., N. C.	+ 50
12. Jackson, Northampton Co., N. C.	− 60
13. Halifax, Halifax Co., N. C.	+ 60
14. Enfield, Halifax Co., N. C.	− 9
15. Tillery, Halifax Co., N. C.	− 130
16. Battleboro, Halifax Co., N. C.	− 8
17. Scotland Neck, Halifax Co., N. C.	− 255
18. Windsor, Bertie Co., N. C.	− 500
19. Tarboro, Edgecomb Co., N. C.	− 275
20. Pinetops, Edgecomb Co., N. C.	− 242
21. Wilson, Wilson Co., N. C.	− 122
22. 9 miles east Wilson	− 112
23. Fountain, Pitt Co., N. C.	− 188
24. Farmville, N. C.	− 385
25. Selma, Johnston Co., N. C.	− 150
26. Pinelevel, Johnston Co., N. C.	+ 120
27. Smithfield, Johnston Co., N. C.	+ 120
28. Freemont, Wayne Co., N. C.	+ 40
29. Goldsboro, Wayne Co., N. C.	− 116
30. Seymour Johnson Field Wayne Co., N. C.	− 120
31. Dunn, Harnett Co., N. C.	+ 120

32. Manchester, Cumberland Co., N. C.	+ 50
33. Fayetteville, Cumberland Co., N. C.	− 100
34. Hamlet, Richmond Co., N. C.	− 125
35. Maxton Glider School, Scotland Co., N. C.	− 143
36. Red Springs, Robeson Co., N. C.	+ 70
37. Chadbourn, Columbus Co., N. C.	− 385
38. Pamlico Sound, Dare Co., N. C. (Esso No. 2)	− 6410a
39. Cape Hatteras, Dare Co., N. C. (Esso No. 1)	− 9960
40. Pamlico, Pamlico Co., N. C. (Atlas Plywood)	− 3406
41. 1 mile S.W. of Pamlico (N. C. Pulpwood)	− 3647
42. Phillippsburg, Carteret Co., N. C. (Wallace)	− 4001
43. Near Merrimon, Carteret Co., N. C. (N. Carraway)	− 4105
44. Near Merrimon (Phillips)	− 3933
45. Near Merrimon (G. Carraway)	− 4038
46. Near Merrimon (Salter)	− 3941
47. Havelock, Craven Co., N. C. (Great Lakes)	− 2288
48. 1 Mile E. Lake Ellis, Craven Co. (Bryan)	− 2361
49. Morehead City, Carteret Co., N. C. (Karsten-Laughton)	− 4019
51. Wilmington, New Hanover Co., N. C.	− 1100
52. Fort Caswell, New Hanover Co., N. C.	− 1530
53. Near Conway, Horry Co., S. C.	− 1369
54. Summerville, Dorchester Co., S. C.	− 2379
55. Florence, Florence Co., S. C.	− 1993a
56. Charleston, Charleston Co., S. C.	− 2005a
57. Parris Island, Beaufort Co., S. C.	− 3436a

a = Total depth not to basement.

Data used in constructing figure 10. Depths are in feet in reference to sea level. Data from Berry (1948), Mundorff (1944), and Richards (1945, 1948).

amount of sand occurs in the Pleistocene formations where small pits are numerous. The sand is used chiefly for building road surfacing and paving, the Highway Department probably using the greater bulk of the material. Concrete Building Block is currently made from these sands.

A moulding sand is derived from pits in the Tuscaloosa formation near Aberdeen, Moore County, and Selma, Johnson County. Furthermore, considerable sand is obtained from the gravel pits in Harnett, Anson Counties, and elsewhere, but here the sand is a secondary product.

GRAVEL

Gravel is less abundant than sand The coarsest gravel occurs in the deposits of Pliocene formation near the "Fall Line" especially in Anson County. A limited amount of finer gravel occurs in the Coharie formation while the amount of gravel in the other Pleistocene formation is negligible. The weathered phase of the Eocene Black Mingo formation exposed in a pit near Lillington consists of gravel which is indistinguishable from the overlying Pliocene.

A complete discussion of the sand and gravel resources of North Carolina was given by Bryson (1930: 102–124). Particular emphasis was paid to the deposits referred to the "Lafayette" formations. A later, but less complete, report was given by Bryson in 1937 (pp. 97–99).

CLAY

KAOLIN

Lenses of pure white kaolin are frequent throughout the Tuscaloosa but they are rarely of economic value.

The finest pure white kaolin found in the present survey came from Chalk Hill, 4 miles southwest of Olivia, Harnett County. Here the kaolin occurred in balls buried about a foot or more in sand on top of the hill. The clay had apparently been reworked from the Tuscaloosa formation.

BRICK CLAY

Brick clays are distributed throughout the state, there being small pits in practically every county. A list of the clay pits operating in 1935 was given by Bryson (1937: 58).

BLEACHING CLAY

Two deposits of bleaching clay of undetermined value have been visited during the present survey. Both clays resemble Fuller's Earth, but they apparently are not that material. Detailed analyses have not yet been made. The localities are listed below:

Location	Formation	No.
Sprunt Farm, 4 m. NE Spout Springs, Harnett Co.	Black Mingo	2
Cannaday Farm, 12 miles below Kinston, Lenoir Co.	Castle Hayne	10

For a discussion of bleaching clays see Nutting (1933).

MARL

The agricultural value of marl as a hard conditioner fertilizer has been known for many years. In fact, much of the early agricultural development in North Carolina was due to the local sources of marl.

Marl may be defined as a calcareous or glauconitic clay or sand. However, the word "marl" has been used loosely to cover many varieties of weathered materials. In this report the term will be used in the first sense defined above. In practically all cases the marl in North Carolina is calcareous; a brief discussion of glauconite will be given in a later section.

It is not the purpose of this report to give a technical discussion of the various marls. Such accounts have been given by Carpenter (1894) and Loughlin, Berry, and Cushman (1921). This report aims to summarize the data on the distribution of marl, the location of the known pits, and possible future reserve supplies.

Before the Civil War, when labor was plentiful, marl was dug on a great many farms in eastern North Carolina. With the changed conditions after 1865, marl digging was less practiced and the farmer began to rely more and more upon commercial land plasters and fertilizers. By the time the field work for Volume 3 was being undertaken (1909–1911), a great many of the old marl pits were idle and many were completely obliterated by fill or water.

Marl is found in various geological formations of the North Carolina Coastal Plain.

CRETACEOUS (BLACK CREEK AND PEEDEE)

Although fossils occur in both of these formations and there is a certain amount of lime, the content and concentration is not sufficiently high to be of commercial or even of significant local agricultural value. No marl pits are known from the Cretaceous at the present time.

Prior to 1928 the Acme Fertilizer Company at Acme, Columbus County, obtained marl from both the Peedee and the overlying Waccamaw.

The glauconitic content of the Peedee formation is discussed in a later section.

EOCENE (CASTLE HAYNE)

This marl is usually well consolidated and not suitable for agricultural purposes without grinding. However, the lime content is very high (usually between 90 and 100 per cent). A somewhat harder phase of the Castle Hayne, such as at Wilmington, Castle Hayne, Maysville, and Belgrade, is used for aggregate and road construction.

MIOCENE (TRENT)

Most of the Trent limestone is too consolidated to be used as agricultural lime although it has been used for building stone aggregate and road construction. Local phases of the Trent near Silverdale consist of unconsolidated shell marl and are used locally as land lime.

A loosely consolidated phase of the Trent formation (intermediate between the loose shell marl and the hard rock) occurs along the Trent River between Pollocksville and New Bern. Several pits or quarries are operated or were formerly operated in this region.

MIOCENE (YORKTOWN)

These marls consist almost entirely of loose shells in sand or clay. They are not as high in lime as the Castle Hayne but are often of considerable local importance. The deposits of Zone 2 are usually higher in lime content than those of the earlier Zone 1.

MIOCENE (DUPLIN)

Unconsolidated shell marl is present in the Duplin formation near Magnolia, Tar Heel, Clinton, Lumberton, and Fairmont. It is very similar to that of the Yorktown formation and is used only locally.

PLIOCENE (WACCAMAW AND CROATAN)

These shell marls are very similar to those of the Yorktown formation and occur mostly in the vicinity of the Cape Fear and Neuse Rivers. Active pits occur near Acme, Elizabethtown, Silverdale, and Padgett.

PLEISTOCENE (PAMLICO)

The shell deposits of the Pamlico formation are almost entirely unconsolidated and so mixed with sand as to render them practically useless for agricultural purposes. However, there is a certain potential value to these Pleistocene shell deposits which are locally so abundant (Hyde County, Dismal Swamp, etc.).

With the bog type of acid soils in the Coastal Plain these marl deposits certainly have a potential value, and with the use of modern machinery they could be excavated in place of the crushed limestone which is hauled long distances from other states.

CEMENT

During the past year or two the possibility of utilizing some of the Coastal Plain limestones for the manufacture of Portland Cement has received consideration. Samples from pits at Belgrade, near Pollocksville (Simmons) and near Trenton (Trent River Limestone and Marl Company) have been analyzed in a preliminary report (Adair, Doody, and Schoenborn, 1947) while the geological possibilities have been discussed by Berry (1947). Satisfactory cements were produced from the limestone having a naturally low silica content and from limestone purified by flotation.

STONE

Consolidated rock occurs rarely in the Coastal Plain although there are a few local deposits of economic importance. Such stone has two chief uses; (1) as building stone and as aggregate and (2) in road construction and surfacing.

By far the most important stone quarry in eastern North Carolina is that of the Superior Stone Company at Belgrade, Onslow County, which has been in operation since 1942 (see fig. 25). The stone comes mainly from the Trent formation and is generally crushed for road material. Other important sources of crushed rock for road material are the Rocky Point Quarry in Pender County, the quarries at Castle Hayne, New Hanover County, and the coquina quarry near Carolina Beach, New Hanover County.

Although not actually a Coastal Plain rock, the now abandoned granite quarries at Wilson and Fountain should perhaps be mentioned in this connection.

GLAUCONITE

During World War I this country experienced a shortage of potash. At that time a careful survey was made of the glauconite deposits of the United States in order to see whether this material could be used as a source of potash. Cretaceous and Eocene deposits in New Jersey and Delaware were found to have the most potential value.

Glauconite had long been known from the Black Creek and Peedee formations of North Carolina, but it was believed that the percentage of potash was too low (Carpenter, 1894). A survey made by Ashley (1917) for the United States Geological Survey showed that one of the best glauconite localities in North Carolina (Edwards Bridge, 6 miles above Grifton, Pitt County) contained 2.46 per cent of potassium and 2.96 per cent of potash. This is contrasted with a typical New Jersey Eocene locality (Birmingham, N. J.) with 5.84 per cent potassium and 7.07 per cent potash.

Even the small amount of glauconite in the Peedee formation gives the sand a noticeable greenish color at various places along the Cape Fear River. Furthermore, the smaller amount of glauconite in the Black Creek formation (Snow Hill stage) gives a greenish color at such places as Snow Hill and Blue Banks Landing.

Glauconitic sands of Cretaceous and Eocene age in New Jersey are used as water softeners.

ILMENITE

Ilmenite sands occur at the bottom of Albemarle Sound and the lower Chowan River. Black ilmenite sand is present in considerable quantities along the south bank of the Chowan River between Black Rock and Mount Gould Landing. It seems most probable that the ilmenite in the Sound has been concentrated from the erosion of the Miocene sediments along the Chowan River. Some outcrops of the ilmenite are slightly consolidated; the so-called "Black Rock" in the river about 100 yards offshore may be of this material.

During the past few years, the DuPont Company has investigated the possibility of dredging the ilmenite from Albemarle Sound, but the operations have now been suspended.

BROMINE

Probably the most unique mineral industry in eastern North Carolina is the bromine plant of the Ethyl-Dow Chemical Company which was in operation until recently at Kure Beach, south of Carolina Beach, New Hanover County. The bromine was extracted from the sea water and used in the manufacture of ethylene dibromide, an important constituent of high grade motor fuel.

PHOSPHATE

Although phosphate formerly occurred commercially in South Carolina, no important deposits have been reported in North Carolina. A limited amount of phosphate nodules can be found in Cretaceous, Eocene, Miocene, and Pliocene deposits in the extreme southern part of the State, but these are not of economic value at the present time. Many years ago an attempt was made to develop the phosphate in the vicinity of Castle Hayne, but this proved unsuccessful.

GROUND WATER

Although perhaps less obvious than some others, ground water is probably the most valuable mineral resource of North Carolina Coastal Plain. Many communities in the eastern part of the State depend entirely upon ground water for their water supply. Except in the case of a few limited areas, the potential supply of ground water under the North Carolina Coastal Plain is enormous and is sufficient to serve the region for many years.

The first systematic survey of the ground water supply of the North Carolina Coastal Plain was made by L. W. Stephenson, B. L. Miller, and B. L. Johnson in 1905 and 1906 and was published as part II of the volume on the Coastal Plain issued by the North Carolina Geological and Economic Survey (Vol. 3).

The United States Geological Survey has prepared reports on the ground water of several local areas in the North Carolina Coastal Plain, but only a few of these (Elizabeth City area, Lohman, 1936, Mundorff 1947; and Halifax area, Mundorff, 1946) have been published.

In 1941, a cooperative program between the North Carolina Geological Survey and the Water Resources Division of the United States Geological Survey was started with the result that systematic records are now being compiled. A list of well logs (Mundorff, 1944) and a Progress Report on ground water (Mundorff, 1945) has been published.

Most of the formations of the Coastal Plain are good aquifers and wells yielding 500 to 1000 gallons of water per minute are not uncommon. According to Mundorff (1945: 7)

soft water can be obtained in approximately the western half of the Coastal Plain, whereas only moderately hard to

TABLE 9

Oil Tests Drilled in North Carolina

Date	County	Location	Company and Name of Well	Elevation	Total Depth	Depth to Basement
1925	Craven	Lake Ellis, Havelock	Great Lakes Drilling Company	30	2,318	2,318
1927	Sampson	Near Clinton	——?——	?	600	—
1945	Carteret	Near Morehead City	Karsten-Laughton No. 1	17	4,044	4,036
1946	Dare	Hatteras Light	Standard Oil of New Jersey—Esso No. 1	24	10,054	9,878
1946	Carteret	½ m. N. Merrimon	Carolina Petroleum Co.—Guy Carraway	15	4,069	4,054
1946	Carteret	2 m. SE. Merrimon	Carolina Petroleum Co.—Anita Carraway	15	4,126	4,120
1946	Carteret	1½ m. N. Merrimon.	Carolina Petroleum Co.—Phillips State	13	3,964	3,946
1946	Carteret	1¼ m. N. Merrimon	Carolina Petroleum Co.—H. B. Salter	13	3,963	3,954
1946	Carteret	1 m. E. Phillips—State well	Carolina Petroleum Co.—John Wallace	11	4,024	4,015
1947	Craven	1 m. S. Ellis Lake	Carolina Petroleum Co.—Bryan	21	2,435	2,408
1947	Dare	Pamlico Sound 11 m. S. Wanchese	Standard Oil of New Jersey—Esso No. 2	21	6,410	—
1947	Pamlico	1 m. SW. Pamlico	Carolina Petroleum Co.—N. C. Pulpwood No. 1	11	3,667	3,658
1947	Pamlico	1½ m. NE. N. C. Pulpwood No. 1 well	Carolina Petroleum Co.—Atlas Plywood No. 1	11	3,430	3,414
1947	Pamlico	1 m. E. Merritt	Carolina Petroleum Co.—Lindley No. 1	16	2,897	?
1949	Hertford	4.8 SE. Cofield	Pam-Beau Drilling Co.—Basemore No. 1	—	1,200	1,175

very hard water can be obtained in most of the eastern half. Brackish water is encountered in many places in the eastern half of the Coastal Plain, at depths ranging from 150 to 500 feet.

PETROLEUM

No petroleum has been obtained from North Carolina. The few wells drilled so far have largely been of a wildcat nature and have produced neither oil nor gas.[7] The most significant wells are the two of the Standard Oil Company of New Jersey in Dare County. The drilling of these two wells was preceded by careful geological and geophysical surveys, and their failure to obtain oil or gas has had a deterring effect on other companies which might have drilled in the State.

Table 9 summarizes the drilling activity in North Carolina.

For further data see Mansfield (1925), Richards (1945a, 1947a, 1947b, 1948a, 1948b, 1949), Stuckey (1949), Straley and Richards (1949).

While no positive traces of oil or gas have been reported from the Coastal Plain of North Carolina, it is still too early to rule out the area as a possible future source of oil or gas (fig. 54, 57, 58).

CORRELATION OF NORTH CAROLINA COASTAL PLAIN FORMATIONS

Brief correlations have been given in the discussions of the various North Carolina formations. Table 10 summarizes and expands these correlations and shows the probable relationship of the various North Carolina formations with deposits in other parts of the Atlantic Coastal Plain. While the table does not include every formation that has been described from the Coastal Plain, it is thought that the major divisions of the stratigraphic divisions of the Atlantic Coastal Plain are mentioned. For further details of correlation see Cooke (1936) and the two correlation tables prepared by the National Research Council in cooperation with the Geological Society of America (Stephenson et al., 1942, Cretaceous) and Cooke, Gardner and Woodring (1943, Cenozoic), as well as the present writer's discussion of the latter (Richards, 1945). The present correlation is adapted from one proposed by the writer in a previous paper (1945a). Attention is also called to table 2 of the present report (page 3) in which is given a comparison of the present interpretation of the stratigraphy of the Coastal Plain of North Carolina with that used in Volume 3 in 1912.

HISTORICAL GEOLOGY

PRE-CAMBRIAN ERA

The beginning of the geological history of North Carolina, as is the case in all parts of the world, is surrounded in mystery. At the earliest stage that can be deciphered from the rocks, it is probable that a sea covered most of the Eastern United States, and some sediments of limestone and carbonaceous material were laid down. It appears unlikely that there was much animal life at this stage. It is also probable that the sea did not cover the region continuously during this era, and that there were periods of erosion when the sea withdrew far to the east.

During some of these periods of erosion, the sedimentary rocks were greatly folded and intruded by igneous or volcanic rocks from the interior of the earth. In terms of the geologist, the layers of sedimentary rocks laid down in the sea were metamorphosed, or changed into schists, marbles, gneisses, etc. At the same time granitic rocks were intruded from the interior of the earth.

Metamorphic rocks of Pre-Cambrian age occur extensively in western North Carolina.

Some, but not all, of the granite occurring in central and western North Carolina is of Pre-Cambrian age. Furthermore, the granite recently encountered at the bottom of the deep oil tests in the eastern part of the State is probably also of Pre-Cambrian age.

By the end of Pre-Cambrian time, there was a prominent land mass made up of these granites, schists, etc., all along eastern North America from New England to Florida. It extended westward into the present Piedmont Plateau and eastward for an unknown distance beyond the present shore line. This land mass is spoken of as *Appalachia* (not to be confused with the Appalachian Mountains). The outcrops of granite and metamorphic rocks near the present "Fall Line" (boundary between the Coastal Plain and the Piedmont Plateau) are remnants of this once extensive mountainous area.

At the "Fall Line," the old surfaces of these ancient granites, gneisses, etc. (frequently called basement rocks) now dip sharply beneath the overlying, mostly unconsolidated Coastal Plain deposits that are of much later age. The entire eastern portion of Appalachia has been eroded and is now far below sea level, and consequently covered with later deposits. This basement rock is encountered in numerous wells at varying depths:

Havelock, Craven County .... 2,318 feet
Merrimon, Carteret County .... 4,053 feet
Morehead City, Carteret County (near) ...... 4,036 feet
Cape Hatteras, Dare County .... 9,878 feet

PALEOZOIC ERA

At the beginning of the Paleozoic Era (Cambrian Period) the low land immediately to the west of the Appalachia was invaded from the southwest by an inland sea which covered a narrow area from the Gulf of Mexico to the Gulf of St. Lawrence, approximately in the position of the present Appalachian Mountains. This early Paleozoic sea covered western North Caro-

---

[7] Slight odors of gas and oil and ether cuts of oil and fluorescence have been reported from some of the wells of the Carolina Petroleum Company near Merrimon from several horizons, but these reported showings have not been proved beyond question to be petroleum.

TABLE 10

Correlation Table of Atlantic Coastal Plain Formations

	NEW JERSEY	DELAWARE	MARYLAND	VIRGINIA	NORTH CAROLINA	SOUTH CAROLINA	GEORGIA	FLORIDA	GULF
PLEISTOCENE	Pensauken, Bridgeton	Cape May; Columbia gr. {Penholoway?, Wicomico?, Sunderland, Brandywine[1]}	Pamlico, Talbot; Penholoway, Wicomico, Sunderland, Coharie, Brandywine?	Pamlico, Talbot; Penholoway, Wicomico, Sunderland, Coharie, Brandywine?	Pamlico, Talbot; Columbia gr. {Penholoway, Wicomico, Sunderland, Coharie, Brandywine?}	Pamlico, Talbot; Penholoway, Wicomico, Sunderland, Coharie, Brandywine	Pamlico, Talbot; Penholoway, Wicomico, Sunderland, Coharie, Brandywine?	Anastasia, Miami, Key Largo, Fort Thompson; ?	Beaumont; Lisle; Willis
PLIOCENE	Beacon Hill	Bryn Mawr	"Lafayette"[2]	"Lafayette"	"Lafayette" Croatan Waccamaw	"Lafayette" Waccamaw	"Lafayette"? Charlton	Caloosahatchee	Goliad
MIOCENE	Kirkwood {Cohansey, St. Marys, Calvert}	Cohansey, St. Marys, Choptank, Calvert	Chesapeake gr. {St. Marys, Choptank, Calvert}	Chesapeake gr. {Yorktown, St. Marys, Choptank, Calvert}	Duplin, Yorktown; Trent	Duplin, Raysor; Hawthorne	Duplin; Hawthorne; Tampa	Choctawatchee; Alum Bluff, Hawthorne; Tampa	Pascagoula—Hattiesburg; Catahoula
OLIGOCENE	—	*	*	Chickahominy	*	Cooper, Santee; Flint River	Cooper, Barnwell, Twiggs cl.; Suwannee	Suwannee, Byram; Ocala	Chickasawhay, Vicksburg; Jackson
EOCENE	Shark River, Manasquan	Pamunkey {Nanjemoy}	Nanjemoy	Nanjemoy	**	McBean	McBean	Avon Park*, Tallahassee*, Lake City*	Claiborne
PALEOCENE	Rancocas {Vincentown, Hornerstown}, Hornerstown (part)*	Aquia; *	Aquia; *	Aquia; *		Black Mingo; *	Wilcox; Clayton	Oldsmar*; Ceder Keys*	Wilcox; Midway
UPPER CRETACEOUS	Monmouth {Tinton, Red Bank, Navesink, Mt. Laurel}, Matawan {Wenonah, Marshalltown, Englishtown, Woodbury, Merchantville}, Magothy, Raritan	Monmouth; Marshalltown, Englishtown, Crosswicks; Magothy; Raritan	Monmouth; Matawan; Magothy; Raritan	—; —; *; *	Peedee; Snow Hill; Black Creek; Black Creek (part)	Peedee; Black Creek; *?	Providence, Ripley; Cusseta, Ripley; Eutaw; Tuscaloosa	*; *; —*; *	Navarro; Taylor; Eutaw; Austin Tuscaloosa
LOWER CRETACEOUS	Potomac* {Patapsco, Arundel*, Patuxent}	Potomac {Patapsco, Arundel*, Patuxent}	Patapsco, Arundel, Patuxent	Patuxent	*—	—	*	*	Woodbine; Comanche

* Subsurface only.
[1] The Brandywine formation is now regarded as Pliocene (Cooke).
[2] "High Level Gravels."
[3] Outcrops near Garner, Clayton, Raleigh, Lillington and in wells near Williamston and elsewhere.

lina, and deposits laid down in this sea at that time are present today in the form of limestone in northeastern McDowell County and vicinity.

This sea existed for millions of years, and occasionally advanced and retreated over surrounding areas, and at times may have withdrawn completely from the trough. Deposits of Middle Paleozoic age (Ordovician, Silurian, Devonian, and Carboniferous) are not known from North Carolina, although they exist farther west in Tennessee. Shellfish, trilobites, corals, fish, and many other marine animals lived in this inland sea and their fossil remains today can be found in rocks which now form the Appalachian Mountains from Alabama to Newfoundland.

In the meantime, the high mountainous area of Appalachia was a prominent feature along the present eastern seaboard and probably covered all of the present eastern North Carolina. Appalachia was uplifted on several occasions during the Paleozoic Era and consequently rapidly flowing rivers carried great quantities of sediment westward into the Paleozoic inland sea. The great weight of these sediments, probably as much as 25,000 feet thick, caused the floor of the inland sea to sink further, forming a great trough or geosyncline.

As the end of the Paleozoic era approached, the inland sea became much shallower, largely because of the sediments continually carried into it from Appalachia. For long periods of time (in the Carboniferous Period) the sea withdrew entirely from the Appalachian region, and in its place immense swamps appeared. The accumulated plants of these swamps were the source of the coal which is so abundant in the Carboniferous formations of Pennsylvania, West Virginia, and Virginia. The sea, however, was never far distant and frequently invaded the swamps, thus explaining the existence of marine limestones or shales interbedded with the coal layers.

No coal deposits of Carboniferous age exist in the State of North Carolina, although it is entirely possible that swamps did exist in western North Carolina during the Carboniferous period and the deposits have since been entirely removed by weathering and erosion.

Sometime during the Carboniferous period granite was intruded from deep within the earth and came to rest beneath the surface. The granites that were intruded at this time and now exposed by erosion are very difficult to distinguish from those that had appeared millions of years earlier, in the Pre-Cambrian. However, many geologists believe that much of the granite and diorite in the vicinity of Raleigh and Durham was intruded in late Paleozoic time.

During the closing periods of the Paleozoic Era (Permian Period) there was an extensive period of uplift and mountain formation, during which time the sediments of the inland sea were pushed upward to form the present Appalachian Mountains in what had been the basin of the inland sea. The rocks that had been laid down in this sea were greatly folded and compressed in the form of majestic mountains.

In the meantime, the land mass of Appalachia had eroded to within a few hundred feet of sea level. Conditions were exactly the reverse of what they had been during the Early Paleozoic. Western North Carolina, which had been low and frequently covered by an inland sea, was now high and mountainous, while the eastern part of the State which had been high (Appalachia) was now eroded down to near sea level.

MESOZOIC ERA

TRIASSIC PERIOD

During the early part of the Mesozoic Era, it is probable that eastern North Carolina was a low lying area, but no evidence can be found to show that it was covered by the sea. The new Appalachian Mountains lay to the west, and were exceedingly rugged, probably as much as 15,000 feet high. Somewhere to the east lay the Atlantic Ocean.

Towards the close of Triassic time, great cracks appeared in the earth's surface in a number of places between Nova Scotia and South Carolina, and long blocks of the earth's crust slumped down thousands of feet, causing a series of basins and depressions. These basins gradually filled with deposits of mud and sand washed from the surrounding highlands. These sediments have now hardened to sandstones and shales. Their gray and reddish-brown colors are characteristic of land basins.

Several such basins existed in North Carolina. One, known as the "Dan River Basin," extended southeastward from Virginia into Rockingham, Stokes and Forsyth Counties. A somewhat similar basin known as the "Deep River Basin" existed some 60 miles to the east from central Granville County to Richmond County, and still a smaller one existed from southern Montgomery County through Anson County into South Carolina.

Recently an additional small Triassic area in North Carolina was described in Davie County (Brown, 1932). The rocks formed in these Triassic basins frequently contain the fossil remains of plants and land animals as well as numerous dinosaur tracks. Coal has been mined in large quantities from the Triassic rocks of the Deep River basin near Cumnock in Chatham and Lee Counties. Thin seams occur in the Dan River Basin.

During Triassic time there was volcanic activity in the region. Conspicuous trap dikes and other igneous rock intrusions are known to cut the Triassic and other older deposits of Eastern America. They are probably most conspicuous in New Jersey where they form the Palisades of the Hudson River, but they are also common in North Carolina.

Toward the end of Triassic time, all rocks laid down during the period were tilted and faulted, causing their present angular position.

JURASSIC PERIOD

The Jurassic period was largely a time of erosion in Eastern America. The Appalachian Mountains began to be weathered and sediments were carried eastward to the sea. As far as we know, the sea covered only extreme eastern North Carolina (near Cape Hatteras).

EARLY CRETACEOUS PERIOD

At the beginning of Lower Cretaceous time the Appalachian Mountains were probably again uplifted with the result that stream erosion increased to a considerable extent. At the same time the present Coastal Plain was slightly tilted downwards toward the southeast. The eroded material was carried eastward by rivers to the Lower Cretaceous sea. Deposits formed by these rivers constitute the Potomac group of formations in Maryland and Virginia, and possibly part of the Tuscaloosa formation of North Carolina.

The Lower Cretaceous shore line was apparently not as far east as that of Jurassic time and the seas covered the extreme eastern part of the state; marine fossils of this age, especially foraminifera and ostracoda, have been found in samples from deep wells in Dare and Carteret Counties.

LATE CRETACEOUS PERIOD

Tuscaloosa Formation

There is probably not a significant break or unconformity between the deposits of the Lower Cretaceous and those of the Upper Cretaceous in eastern North Carolina. Erosion from the mountains continued with the result that sediments continued to be carried down toward the sea. The sands and clays of the Tuscaloosa formation that crop out in North and South Carolina are thought to have been carried down from the highlands and deposited in swamps or estuaries. No fossils have yet been found from the outcrops of the Tuscaloosa formation in central North Carolina, although a number of plant fossils have been found in South Carolina, especially near Middendorf, in Chesterfield County.

However, the sea did cover extreme northeastern North Carolina during Tuscaloosa time, as shown by the finding of marine fossils in well samples at Merrimon and Morehead City in Carteret County, and near Cape Hatteras, in Dare County, as well as near Norfolk, Va.

Black Creek Formation

The sea apparently withdrew at the close of Tuscaloosa time, and for a while there were extensive forests and swamps covering much of the Coastal Plain of North and South Carolina. These swamp deposits constitute the early or "continental" phase of the Black Creek formation. Fossil wood is abundant at many places, for example, near Smithfield, Johnston County (fig. 15), Fayetteville, Cumberland County, and at many localities in South Carolina.

Far to the east, these swamps joined the sea, and marine deposits of this age (Early Black Creek = Eutaw or Magothy) have been recognized in well samples at Hatteras and elsewhere.

The sea began to creep inland, engulfing the vegetation and by the middle of Black Creek time covered most of the Coastal Plain of North Carolina.

This Upper Cretaceous sea deposited dark colored sands and laminated clays, and later alternating beds of sands and clays. Fossil shells are found throughout the formation, but these are more common in the upper part which is known as the Snow Hill member, named from rich deposits at Snow Hill in Greene County.

Peedee Formation

There was apparently no break, or complete withdrawal of the sea in Eastern North Carolina between the deposition of the Black Creek formation and that of the Peedee formation. However, farther south, in South Carolina the lower part of the Peedee formation is missing, indicating a temporary emergence of the area at the end of Black Creek time.

During the deposition of the Peedee formation, the sea probably did not extend quite as far inland over the North Carolina Coastal Plain as it had during Black Creek time. The deposits formed in this sea include interbedded sands, clays, and marls.

Marine life was abundant during the latter part of Upper Cretaceous time as indicated by the presence of fossils. Land animals were probably also conspicuous, although they have left very few fossil remains here. Dinosaurs and other large reptiles roamed the lands while the swamps and rivers contained Mosasours and other large crocodile-like animals.

CENOZOIC ERA

EOCENE PERIOD

Early and Middle Eocene Time

At the close of the Cretaceous time the seas withdrew toward the east from the Coastal Plain and there followed a long period of erosion. This withdrawal of the sea was probably caused by an uplift of the land, although the details are not entirely clear.

Farther south, in Georgia and along the coast of the Gulf of Mexico, the sea advanced and covered parts of the Coastal Plain during Early Eocene time (Midway), but there is no evidence that this sea covered any parts of North Carolina.

When the sea next advanced over North Carolina, it spread inland as far as the "Fall Line," and even beyond. Deposits referred to the Middle Eocene occur in Harnett, Hoke, Wake, and Johnston Counties. The farthest inland that fossiliferous deposits of this age have been found is at a point in Crabtree Park in Wake County between Raleigh and Durham. Because of the scattered nature of these deposits, it has been impossible to date them more closely than to say that they are of Middle Eocene age, and that possibly they may be correlated with the Claiborne formation of the Gulf Coast.

Other fossils referred to the Middle Eocene have been taken from wells at Williamston, Martin County, and elsewhere in eastern and northeastern North Carolina, so it is probable that this sea covered all of the Coastal Plain of the state.

The Lower and Middle Eocene seas covered much of the Virginia Coastal Plain where they formed the Aquia and Nanjemoy formations (Pamunkey group).

Castle Hayne Formation

Late Eocene time (Jackson time) may have begun with an uplift of the land that raised the region between the Cape Fear River in North Carolina and the Santee River in South Carolina. On the other hand, it is quite possible that this uplift took place during an earlier period. This uplift, which existed for millions of years until the Middle Miocene, is spoken of as the "Great Carolina Ridge" or the "Cape Fear Uplift." The sea invaded the land both north and south of this ridge. South of the ridge the formation is named the Santee limestone, while north of it, in North Carolina, it has been called the Castle Hayne formation.

The Castle Hayne sea submerged at least the eastern portion of the North Carolina Coastal Plain, although it is probable that it did not extend quite as far inland, as had the seas of Middle Eocene time Limestones were the prevailing deposits of this sea. Fossiliferous outcrops of this formation occur in many places, especially near Wilmington, Castle Hayne, and Dover.

OLIGOCENE PERIOD

Very little is known about the Oligocene history of North Carolina. The recent discovery of Oligocene microfossils in wells at Camp Lejeune and Cape Hatteras suggests that only the extreme eastern part of the state was covered by the sea during this period.

MIOCENE PERIOD

Trent Formation

The "Great Carolina Ridge" remained out of water during Early and Middle Miocene time. There was probably a long period of erosion after the deposition of any Oligocene material which accounts for the fact that most of the sediments of that age have been removed by erosion.

The first marine invasion of North Carolina during Miocene Time was a relatively slight embayment and appears to have submerged only Onslow and Jones Counties and the land east of them. The material deposited in this sea is known as the Trent formation, and consists largely of limestone, rather similar to that deposited during Castle Hayne time, although usually somewhat more consolidated and tougher. The best fossil localities occur near Belgrade and Silverdale in Onslow County.

Possibly the Trent sea extended over the present land north of Cape Hatteras but, if so, the deposits have long since been eroded, and it seems more probable that this invasion was more or less local.

A contemporaneous invasion of the sea occurred far to the south, in southern Georgia and Florida where it formed the Tampa limestone.

Calvert Formation

The Trent sea withdrew and there followed a period of erosion. Then the land depressed again, but not uniformly. The depression was greater to the north with the result that the next sea (Calvert) covered much of the Coastal Plain of New Jersey, Delaware, Maryland, and Virginia, but probably only the extreme northeastern part of North Carolina. Probably eastern North Carolina was relatively higher at this time so that the sea did not advance very far inland.

St. Mary's Formation

As a result of an uplift of the land, the Calvert sea withdrew, probably far beyond the present shore line. There followed another depression of the land, but again this depression was greatest along the Coastal Plain north of North Carolina. No deposits that can be positively referred to this invasion—the St. Mary's—are known from North Carolina.

Yorktown Formation

The next time that there was a sinking of the land along the Atlantic Coast, it was much more extensive and extended from Virginia to Florida. By this time the "Great Carolina Ridge" had subsided, allowing the sea to cover most of the Coastal Plain of North and South Carolina. The formation laid down by this sea is called the Yorktown formation, and is extensively developed in eastern North Carolina. The climate was warm, as demonstrated by the rich fauna of marine mollusks and other invertebrates. Whales and sharks were also abundant in the Yorktown sea and their fossils, especially shark teeth, are common in many localities.

The Duplin marl, which is especially conspicuous in southeastern North Carolina and in South Carolina, was deposited during a late phase of Yorktown time. The climate at this stage was probably very mild and there was an unusual abundance of marine life. Much of this formation today consists of concentrations of friable shells which are good for agricultural uses.

PLIOCENE PERIOD

Waccamaw Formation

There probably was a slight withdrawal of the sea at the close of Miocene time, although this is not certain. The sea advanced again, owing to another sinking of the land, but this time the sinking was more pronounced in the southern part of the state, and farther south.

The early Pliocene shore line probably lay some 40 miles inland from the present shore position in South Carolina. It extended northeastward across North Carolina and probably cut the present shore line near the Virginia Capes. No marine Pliocene deposits are known north of North Carolina. Two formations have been described from the early Pliocene of North Carolina—the Waccamaw and the Croatan—but they are probably contemporaneous. Fossils are abundant, suggesting a warm sea.

The close similarity between the fauna of the Waccamaw and that of the Duplin (Miocene) suggests that there was only a slight time interval, or none at all, between Miocene and Pliocene time in North Carolina.

Late Pliocene Time

The land was high during the remainder of Pliocene time. Possibly there was a renewed uplift of the Appalachian region. At any rate, rivers carried extensive deposits of gravel and spread them out on the Coastal Plain, especially near the contact with the Piedmont Plateau (Fall Zone). These gravels extend south and southwest across the state roughly from Northampton and Halifax Counties to Richmond and Anson Counties at the South Carolina border. No fossils have been found in these gravels and some geologists believe that they date from the Early Pleistocene.

PLEISTOCENE PERIOD

Although the Pleistocene is generally spoken of as the Ice Age, it is known that the climate was not continuously cold. The ice advanced several times, probably four, but these glacial stages were separated from each other by periods of time, known as interglacial stages, when the climate was at least as warm as today, and probably considerably warmer.

The glaciers of the Pleistocene did not advance farther south than New Jersey and therefore North Carolina had no glaciation, although the climate was undoubtedly cold during the times of glacial advance.

The history of the Early and Middle Pleistocene in North Carolina is not clear, and the evidence is subject to different interpretations.

According to some geologists (especially Cooke) the sea covered the Coastal Plain during the various interglacial stages. At these times there was more water in the sea because of the melting of the continental ice sheets with the result that sea level was considerably

higher than at present. Conversely, sea level dropped to below its present level when water was taken from the ocean for the formation of huge ice masses during the times of glacial advance. According to Cooke, the sea was 270 feet higher than at present during the first interglacial stage, and somewhat less high during each of the succeeding interglacial stages. The levels are marked by so-called "Terraces."

According to the alternate interpretation (especially Flint), only the youngest (lowest) two "Terraces" are of marine origin, the others having been formed by rivers. This theory does not question the fact that the rise and fall of sea level is governed by the glaciers, but questions whether the sea actually rose as much as 270 feet in order to form the highest terraces as described by Cooke.

Favoring this second interpretation is the complete absence of marine fossils, other than in the lowest terrace deposit (Pamlico formation), and the scarcity or absence of marine features above the 40 foot elevation.

Early Pleistocene Time

The Early Pleistocene history of the Coastal Plain has yet to be completely deciphered. Undoubtedly, sea level changes took place during the advances and retreats of the ice, but possibly movements of the land have complicated the picture so that it is difficult to interpret the story correctly.

Late Pleistocene Time

It is a little easier to interpret the geological history of North Carolina during the latter part of the Pleistocene. During the last major interglacial stage sea level was about 25 feet higher than at present. This interglacial shore line can be traced all the way from northern New Jersey to Florida and along the Gulf Coast.

A rise of sea level of 25 feet would have submerged Cape Hatteras. The result would have been that the Gulf Stream, instead of being deflected out to sea at this Cape, as is the case today, would have remained close to shore as far north as Cape Cod. The presence of warm climate mollusks and other fossils at many localities along the East Coast as far north as Nantucket Island, Massachusetts, supports this view.

The presence of this warm interglacial sea in eastern North Carolina is clearly shown by fossils which underlie that part of the region that has an elevation of 25 feet or less. Such fossils occur along the Neuse River below New Bern, and have been dredged in great numbers from the excavations for the Intracoastal Waterway canal in Hyde and Pamlico Counties.

As the glaciers grew and advanced from the north, sea level fell, and the climate became cooler. Evidences of this glacial shore line are difficult to find, because at the climax of the last glaciation (Wisconsin Time) sea level was probably about 300 feet lower and the shore line was some 90 miles east of the present beach. However, there may have been bays and inlets during this low sea level stage. Very recently some fossil shells of cold water species were dredged at Manteo, North Carolina (see p. 41), and it is thought that these may have lived in a bay or inlet during a glacial stage.

Another indication of a cold glacial climate is the finding of fossil bones of the walrus on the beach at Kitty Hawk.

RECENT TIME

As the ice sheets withdrew and melted, water was freed and sea level again began to rise, and the shore lines advanced to their present position. It is probably still rising, but at an almost imperceptible rate. Old sod and tree stumps in places are exposed along the barrier beaches near Cape Hatteras, North Carolina, Myrtle Beach, South Carolina, and elsewhere, and clearly indicate that the land is receding or the seas encroaching. How much of this loss of land is due to a rise in sea level and how much to the action of waves and currents is problematical. Probably both factors are involved.

BIBLIOGRAPHY

Volume 3 of the North Carolina Geological and Economic Survey published in 1912 contained a complete bibliography of all references on the Coastal Plain of North Carolina from 1791 to 1912. The present bibliography is in two parts:

A. List of references on the geology and paleontology of North Carolina published since 1912; also a few references of slightly earlier date that were omitted from the bibliography published in 1912.

B. Additional references cited in the text of this report including some published prior to 1912 as well as some more recent works dealing largely with regions outside of the North Carolina Coastal Plain.

BIBLIOGRAPHY A

ADAIR, R. B., T. C. DOODY, AND E. M. SCHOENBORN. 1947. Evaluation of North Carolina raw material for manufacture of cement. I. Preliminary laboratory investigation. *N. C. State College Record* **46** (9) (*Dept. of Engineering Research Bull.* 35).

ANTEVS, ERNST. 1929. Quaternary marine terraces in nonglaciated regions and changes of level of sea and land. *Amer. Jour. Sci.* **17** : 35–49.

ASHLEY, GEORGE. 1917. Notes on the greensand deposits of the eastern United States. *U. S. Geol. Surv. Bull.* **660-B**.

BERRY, CHARLES T. 1931. Metatarsal of Equus from marine Pliocene of North Carolina. *Pan Amer. Geol.* **56** : 340–342.

BERRY, E. W. 1910. Geologic relations of the Cretaceous floras of Virginia and North Carolina. *Bull. Geol. Soc. Amer.* **20** : 655–659.

—— 1914. The Upper Cretaceous and Eocene floras of South Carolina and Georgia. *U. S. Geol. Surv. Prof. Paper* **84**.

—— 1919. Upper Cretaceous floras of the eastern Gulf Region in Tennessee, Mississippi, Alabama and Georgia. *U. S. Geol. Surv. Prof. Paper* **112**.

—— 1920. Contributions to the Mesozoic flora of the Atlantic Coastal Plain, XIII : North Carolina. *Torrey Bot. Club Bull.* **47** : 397–406.

—— 1926. Pleistocene plants from North Carolina. *U. S. Geol. Surv. Prof. Paper* **140-C**.

BERRY, E. W., AND J. A. CUSHMAN. 1921. See Loughlin, Berry and Cushman, 1921.

BERRY, WILLARD. 1942. Water supply in North Carolina Coastal Plain. *Bull. Geol. Soc. Amer.* **53** : 1795.

—— 1942a. Pleistocene plants from South Carolina. *Amer. Jour. Botany* **29** : 25 (abstract).

—— 1942b. Geology of three wells in the Coastal Plain of North Carolina. *Jour. Elisha Mitchell Sci. Soc.* **58** : 137.

—— 1943. Recent wells near Elizabeth City, N. C. *Jour. Elisha Mitchell Sci. Soc.* **59** : 118.

—— 1943a. Exogyra costata zone in Horry County, S. C. *Jour. Elisha Mitchell Sci. Soc.* **59** : 118.

—— 1946. New concept of North Carolina Coastal Plain. *Bull. Geol. Soc. Amer.* **57** : 1177–1178.

—— 1948. North Carolina Coastal Plain floor. *Bull. Geol. Soc. Amer.* **59** : 87–89.

—— 1947=1949. Marls and limestones of eastern North Carolina. *N. C. Dept. Consv. & Develop. Bull.* **54**.

—— 1949. Fossils from Harrellsville, North Carolina. *Jour. Elisha Mitchell Sci. Soc.* **65** : 196.

BERRYHILL, LOUISE. 1948. Stratigraphy of Standard Oil of New Jersey's Hatteras Light Well No. 1. *Jour. Elisha Mitchell Sci. Soc.* **64** : 170–171.

BOESHORE, IRWIN, and WILLIAM D. GRAY. 1936. An Upper Cretaceous wood: Torreya antiqua. *Amer. Jour. Bot.* **23** : 524–528.

BOON, J. D., and C. C. ALBRITTON, JR. 1938. Established and supposed examples of meteoric craters and structures. *Field & Lab.* **16** : 44–56.

BROWN, A. P., and H. A. PILSBRY. 1912. Note on a collection of fossils from Wilmington, North Carolina. *Proc. Acad. Nat. Sci. Phila.* **64** : 152–153.

BRYSON, HERMAN J. 1928. The story of the geological making of North Carolina. *N. C. Dept. Consv. and Devl. Educational Series* No. 1.

—— 1930. The mining industry in North Carolina during 1927 and 1928. *N. C. Dept. Consv. and Devl. Econ. Paper* **63**.

—— 1937. The mining industry in North Carolina from 1929 to 1936. *Ibid.* **64**.

BUELL, MURRAY F. 1939. Peat formation in the Carolina Bays. *Bull. Torrey Bot. Club* **66** : 483–487.

—— 1946. The age of the Jerome Bog, a "Carolina Bay." *Science* **103** : 14–15.

CANU, F., and R. S. BASSLER. 1920. North American Early Tertiary Bryozoa. *Bull. U. S. Nat. Mus.* No. 106.

CANU, F., and R. S. BASSLER. 1923. North American later Tertiary and Quaternary Bryozoa. *Bull. U. S. Nat. Mus.* No. 125.

CEDERSTROM, D. J. 1945. Structural geology of southeastern Virginia. *Bull. Amer. Assn. Petr. Geol.* **29** : 71–95.

CLARK, W. B. 1910. Results of a recent investigation of the coastal plain formations in the area between Massachusetts and North Carolina. *Bull. Geol. Soc. Amer.* **20** : 646–654.

CLARK, W. B., A. B. BIBBINS, and E. W. BERRY. 1911. The Lower Cretaceous deposits of Maryland. The Lower Cretaceous floras of the world. *Md. Geol. Surv.*

CLARK, W. B., and M. TWITCHELL. 1915. The Mesozoic and Cenozoic Echinodermata of the United States. *U. S. Geol. Surv. Mon.* **54**.

CLARK, W. B. et al. 1916. Upper Cretaceous. *Md. Geol. Surv.*

COBB, COLLIER. 1923. The immediate ancestor of our domestic horse found fossil in Halifax County, North Carolina. *Jour. Elisha Mitchell Sci. Soc.* **39** : 31–32.

—— 1933. Dune sand and eolian soils in relation to present and past climatic conditions of the continent of North America. *Cong. intern. geographie, Paris, 1931. Compte rendu* **2** (1) : 712.

COBB, WILLIAM B. 1932. Variations in soils developed from sands in eastern North Carolina. *Jour. Elisha Mitchell Sci. Soc.* **27** : 17–18.

COOKE, C. WYTHE. 1916. The age of the Ocala limestone. *U. S. Geol. Surv. Prof. Paper* **95**.

—— 1925. Correlation of the Eocene formations of Mississippi and Alabama. *U. S. Geol. Surv. Prof. Paper* **140-E**.

—— 1930. Correlation of coastal terraces. *Jour. Geol.* **38** : 577–589.

—— 1931. Pleistocene seashores. *Jour. Wash. Acad. Sci.* **20** : 389–395.

—— 1931a. Seven coastal terraces in the southeastern states. *Ibid.* **21** : 503–513.

—— 1932. Tentative correlation of American glacial chronology with the marine time scale. *Ibid.* **22** : 310–312.

—— 1933. Tentative ages of Pleistocene shore lines. *Ibid.* **23** : 331–333.

—— 1934. Discussion of the origin of the supposed meteorite scars of South Carolina. *Jour. Geol.* **42** : 88–104.

—— 1936. Geology of the Coastal Plain of South Carolina. *U. S. Geol. Surv. Bull.* **867**.

—— 1937. The Pleistocene Horry Clay and Pamlico formation near Myrtle Beach, S. C. *Jour. Wash. Acad. Sci.* **27** : 1–5.

—— 1940. Elliptical bays in South Carolina and the shape of eddies. *Jour. Geol.* **48** : 205–211.

—— 1941. Cenozoic regular echinoids of eastern United States. *Jour. Paleont.* **15** : 1–20.

—— 1942. Cenozoic irregular echinoids of eastern United States. *Ibid.*, **16** : 1–62.

—— 1943. Elliptical bays. *Jour. Geol.* **51**: 419–427.

COOKE, C. W., JULIA GARDNER, and W. P. WOODRING. 1943. Correlation of the Cenozoic formations of the Atlantic and Gulf Coastal Plain and the Caribbean region. *Bull. Geol. Soc. Amer.* **54**: 1713–1722.

CUSHMAN, J. A. 1918. Some Pliocene and Miocene Foraminifera of the Coastal Plain of the United States. *U. S. Geol. Surv. Bull.* **676**.

—— 1935. Upper Eocene Foraminifera of the southeastern United States. *U. S. Geol. Surv. Prof. Paper* **181**.

CUSHMAN, J. A., and E. D. CAHILL. 1933. Miocene Foraminifera of the Coastal Plain of the Eastern United States. *U. S. Geol. Surv. Prof. Paper* **175**.

DALY, R. A. 1934. The changing world of the Ice Age. New Haven, Yale Univ. Press.

DAVIS, H. T., and H. G. RICHARDS. 1942. Reviewing North Carolina Coastal Plain geology. *Jour. Elisha Mitchell Sci. Soc.* **58**: 136.

EWING, MAURICE, J. L. WORZEL, N. C. STEENLAND, and F. PRESS. 1946. Geophysical investigations in the emerged and submerged Atlantic Coastal Plain. Part 5. Cape May, New York and Woods Hole Sections. *Bull. Geol. Soc. Amer.* **57**: 1192 (abstract).

EWING, MAURICE, A. P. CRARY, and N. M. RUTHERFORD. 1937. Geophysical investigations in the emerged and submerged Atlantic Coastal Plain. Part 1, Methods and results. *Bull. Geol. Soc. Amer.* **48**: 758–802.

EWING, MAURICE, GEORGE P. WOOLLARD, and A. C. VINE. 1939. Geophysical investigations in the emerged and submerged Atlantic Coastal Plain. Part 3, Barnegat Bay, New Jersey section. *Bull. Geol. Soc. Amer.* **50**: 257–296.

FABIANIC, W. L., and N. H. STOLTE. 1933. Mineralogy of typical North Carolina clays and shales. *Amer. Ceramic Soc. Jour.* **16**: 6–11.

FERGUSON, J. L. 1943. Review of "The Origin of the Carolina Bays" by Douglas Johnson. *Bull. Amer. Assn. Petr. Geol.* **27**: 654–656; 874.

FLINT, R. F. 1940. Pleistocene features of the Atlantic Coastal Plain. *Amer. Jour. Sci.* **238**: 757–787.

—— 1942. Atlantic coastal "terraces." *Jour. Wash Acad. Sci.* **32**: 235–237.

—— 1947. Glacial geology and the Pleistocene epoch. N. Y., John Wiley and Sons.

GARDNER, JULIA. 1915. Relation of the Late Tertiary faunas of the Yorktown and Duplin formations. *Amer. Jour. Sci.*, ser. 4, **39**: 305–310.

—— 1925. Coastal plain and European Miocene and Pliocene Molluska. *Bull. Geol. Soc. Amer.* **35**: 857–866.

—— 1928. A new gastropod from the Miocene of Virginia. *Jour. Wash. Acad. Sci.* **18**: 561–563.

—— 1943. Mollusca from the Miocene and Lower Pliocene of Virginia and North Carolina. Part 1, Pelecypoda. *U. S. Geol. Surv. Prof. Paper* **199-A**.

—— 1948. Mollusca from the Miocene and Lower Pliocene of Virginia and North Carolina. Part 2, Scaphopoda and Gastropoda. *U. S. Geol. Surv. Prof. Paper* **199-B**.

GARDNER, JULIA, and T. H. ALDRICH. 1919. Mollusca from the Upper Miocene of South Carolina with descriptions of new species. *Proc. Acad. Nat. Sci. Phila.* **71**: 17–53.

GISLEN, TORTSEN. 1934. A reconstruction problem; analysis of fossil comatulids from North America, with a survey of all known types of comatulid arm ramifications. *Univ. Arssk. (Acta Univ. Lundensis) neue Folge* **30**: Avd. 2 (K. Fys. Sallsk Lund Handl., neue Fokge, Bank 45), Nr. 11.

GLENN, L. C. 1895. Some notes on Darlington (South Carolina) Bays. *Science* **2**: 472–475.

—— 1938. Geology of the Intracoastal Canal near Myrtle Beach, South Carolina. *Proc. Geol. Soc. Amer. for 1937*: 83.

HARBISON, ANNE. 1944. Mollusks from the Eocene Santee limestone South Carolina. *Acad. Nat. Sci. Phila. Not. Nat.* **143**.

HARRIS, G. D. 1919. A few Mid-Upper Eocene Fossils from the Carolinas and Texas. *Bull. Amer. Paleont.* **8**: 13–18.

—— 1919. Pelecypoda of the St. Maurice and Claiborne stages. *Ibid.* **6** (31).

HAY, O. P. 1923. The Pleistocene of North America and its vertebrated animals from the states east of the Mississippi River and from the Canadian Provinces east of longitude 95. *Carn. Inst. Wash. Pub.* 322.

HAZLETT, DONALD C. 1938. Beach erosion control at Cape Hatteras, North Carolina. *Bull. Geol. Soc. Amer.* **49**: 1950.

HENBEST, L. G., K. E. LOHMAN, and W. C. MANSFIELD. 1939. Foraminifera, diatoms and mollusks from test wells near Elizabeth City, N. C. *U. S. Geol. Surv. Prof. Paper* **189-G**.

HITE, M. P. 1924. Some observations of storm effects on ocean inlets. *Amer. Jour. Sci.*, ser. 5, **7**: 319–326.

HOWELL, B. F. 1943. Hamulus, "Falcula" and other Cretaceous Tubicola of New Jersey. *Proc. Acad. Nat. Sci. Phila.* **95**: 139–166.

HUDDLE, J. W. 1940. Notes on the geological section at the natural well near Magnolia, N. C. *Jour. Elisha Mitchell Sci. Soc.* **56**: 227–228.

INGRAM, WILLIAM M. 1939. A new fossil cowry from North Carolina. *Nautilus* **52**: 120–121.

JOHNSON, C. W. 1904. Description of two new Tertiary fossils. *Nautilus* **17**: 143–144.

JOHNSON, DOUGLAS. 1934. Supposed meteorite scars of South Carolina. *Science* **79**: 461.

—— 1936. Origin of the Carolina Bays. *Proc. Geol. Soc. Amer. for 1935*: 84.

—— 1936a. Origin of the supposed meteorite scars of Carolina. *Science* **84**: 15–18.

—— 1937. "Cicatrices meteoritiques" sur la côte des Carolines. *In* Melanges de Geographie et d'Orientalisme offerts a E. F. Gautier. Tours, France.

—— 1937. Role of artesian waters in forming the Carolina Bays. *Science* **86**: 255–258.

—— 1940. Mysterious craters of the Carolina coast. *In* Science in Progress, New Haven, Society of Sigma XI.

—— 1941. Rotary currents and the Carolina Bays. *Jour. Geomorph.* **4**: 164–166.

—— 1942. The origin of the Carolina Bays. N. Y., Columbia Univ. Press.

—— 1944. Mysterious craters of the Carolina Coast. *Amer. Scientist* **32**: 1–22.

JOHNSON, W. R., JR. 1938. Geomagnetic reconnaissance of the coastal plain of Northeastern North Carolina. *Bull. Geol. Soc. Amer.* **49**: 1951.

KELLUM, L. B. 1925. The age of the Trent marl in North Carolina. *Jour. Geol.* **33**: 183–187.

—— 1926. Paleontology and stratigraphy of the Castle Hayne and Trent marls in North Carolina. *U. S. Geol. Surv. Prof. Paper* **143**.

—— 1931. Revision of the names of three fossils from the Castle Hayne and Trent marls in North Carolina. *Jour. Wash. Acad. Sci.* **21**: 51–52.

KJELLESVIG, ERIK N. 1933. Variations in the test of Nonion pizarrensis Berry and Nonionella auris (D'Orbigny) from the Miocene of North Carolina. *Elisha Mitchell Sci. Soc. Jour.* **49**: 24.

LAHEE, FREDERIC. 1949. Exploratory drilling in North Carolina in 1946. *Bull. Amer. Assn. Petrol. Geol.* **33**: 1903.

LEWIS, IVEY, and E. C. COCKE. 1929. Pollen analysis of Dismal Swamp peat. *Elisha Mitchell Sci. Soc. Jour.* **45**: 37–58.

LOHMAN, S. W. 1936. Geology and ground water resources of the Elizabeth City Area, North Carolina. U.S.G.S. Water Supply Paper 773-A.

LOUGHLIN, G. F., E. W. BERRY, and J. A. CUSHMAN. 1921. Limestones and marls of North Carolina. N. C. Geol. and Econ. Surv. Bull. 28.

MACCARTHY, G. R. 1936. Magnetic anomalies and geologic structures of the Carolina Coastal Plain. Jour. Geol. 44: 396–406.

—— 1936a. The Carolina Bays. Proc. Geol. Soc. Amer. for 1935: 90–91.

—— 1936b. Meteors and the Carolina Bays. Jour. Elisha Mitchell Sci. Soc. 50: 211.

—— 1937. The Carolina Bays. Bull. Geol. Soc. Amer. 48: 1211–1226.

MACCARTHY, G. R., and J. A. ALEXANDER. 1934. What lies under the Coastal Plain? Jour. Elisha Mitchell Sci. Soc. 50: 50.

MACCARTHY, G. R., W. F. PROUTY, and S. A. ALEXANDER. 1933. Some magnetic observations in the Coastal Plain area of South Carolina. Jour. Elisha Mitchell Sci. Soc. 49: 20–21.

MACCARTHY, G. R., and H. W. STRALEY, III. 1938. Geomagnetic reconnaissance of the Carolina Coastal Plain. Bull. Geol. Soc. Amer. 49: 1953.

MCCAMPBELL, JOHN. 1944. An evaluation of the artesian hypothesis of the origin of the Carolina Bays. Jour. Elisha Mitchell Sci. Soc. 60: 183–185.

—— 1945. A geomagnetic survey of some Bladen County, North Carolina, "Carolina Bays." Jour. Geol. 53: 66–67.

MCKELVEY, V. E., and J. R. BALSLEY. 1948. Distribution of coastal sands in North Carolina, South Carolina and Georgia as mapped from an airplane. Econ. Geol. 43: 518–524.

MCLEAN, JAMES. 1947. Oligocene and Lower Miocene microfossils from Onslow County, North Carolina. Acad. Nat. Sci. Phila. Not. Nat. No. 200.

MACNEIL, F. S. 1936. A new Crassatellid from the Waccamaw formation of North and South Carolina and the Caloosahatchee marl of Florida. Jour. Wash. Acad. Sci. 26: 528–530.

—— 1938. Species and genera of Tertiary Noetinae. U. S. Geol. Surv. Prof. Paper 189-A.

—— 1940. Supplemental note of the occurrence of Tertiary Noetinae. Jour. Paleont. 14: 507–509.

MANSFIELD, W. C. 1925. Oil prospecting well near Havelock, N. C. N. C. Dept. Consv. and Devl. Econ. Paper 58.

—— 1928. Notes on Pleistocene faunas from Maryland, Virginia and Pliocene and Pleistocene faunas from North Carolina. U. S. Geol. Surv. Prof. Paper 150-F.

—— 1928a. New fossil mollusks from the Miocene of Virginia, with a brief outline of the divisions of the Chesapeake group. Proc. U. S. Nat. Mus. 74, art. 14: 1–11.

—— 1929. The Chesapeake Miocene basin of sedimentation as expressed in the new geologic map of Virginia. Jour. Wash. Acad. Sci. 19: 263–268.

—— 1936. Additional notes on the Molluscan fauna of the Pliocene Croatan sand of North Carolina. Jour. Paleont. 10: 665–668.

—— 1936a. Stratigraphical significance of Miocene, Pliocene and Pleistocene Pectinidae in the southeastern United States. Jour. Paleont. 10: 168–192.

—— 1937. Some deep wells near the Atlantic Coast in Virginia and the Carolinas. U. S. Geol. Surv. Prof. Paper 186-I.

—— 1937a. Mollusks of the Tampa and Suwannee limestones of Florida. Fla. Dept. Consv. Geol. Bull. 15.

—— 1937b. A new subspecies of Pecten from the Upper Miocene of North Carolina. Jour. Wash. Acad. Sci. 27: 10–12.

MANSFIELD, W. C., and F. S. MACNEIL. 1937. Pliocene and Pleistocene mollusks from the Intra-Coastal Waterway in South Carolina. Ibid. 27: 5–10.

MELTON, F. A. 1934. The origin of the Carolina "Bays." Discovery 15: 151–154.

—— 1934a. The origin of the supposed meteorite scars: Reply. Jour. Geol. 42: 97–104.

—— 1938. Possible Late Cretaceous origin of the Carolina Bays. Bull. Geol. Soc. Amer. 49: 1954.

—— 1950. The Carolina "Bays." Jour. Geol. 58: 128–134.

MELTON, FRANK A., and WILLIAM SCHRIEVER. 1933. The Carolina "Bays"—are they meteor scars? Jour. Geol. 41: 52–66.

MELTON, F. A., and WILLIAM SCHRIEVER. 1933. Meteorite scars in the Carolinas. Bull. Geol. Soc. Amer. 44: 94.

MILLER, B. L. 1937. Geophysical investigations in the emerged and submerged Atlantic Coastal Plain. Part 2, Geological significance of the geophysical data. Bull. Geol. Soc. Amer. 48: 803–812.

MONROE, WATSON H. 1938. Pleistocene shoreline features in southeastern Virginia. Bull. Geol. Soc. Amer. 49: 1938.

MUNDORFF, M. J. 1944. Selected well logs in the coastal plain of North Carolina. N. C. Dept. Consv. and Devl. Information Circular No. 3.

—— 1945. Progress report on ground water in North Carolina. N. C. Dept. Consv. & Devl. Bull. 47.

—— 1946. Ground water in the Halifax area, North Carolina. N. C. Dept. Consv. and Devl. Bull. 51.

—— 1947. A possible new source of ground-water supply in the Elizabeth City area, North Carolina. N. C. Dept. Consv. and Develop. Information Circular 6.

—— 1947a. Ground water in the Neuse River Basin, North Carolina. In Hydrologic data on the Neuse River Basin, 1865–1945, 89–115, N. C. Dept. Consv. and Develop.

—— 1947b. Ground-water in the Cape Fear River Basin, North Carolina. In Hydrologic data on the Cape Fear River Basin, 1820–1945, 127–164. N. C. Dept. Consv. and Develop.

NELSON, WILBUR A. 1940. Topography of the former continent of Appalachia (from geologic evidence). Trans. Amer. Geophys. Union 3A: 786–796.

OLSSON, AXEL A. 1914. Notes on Miocene correlation. Nautilus 27: 101–103.

—— 1914a. New and interesting Neocene fossils from the Atlantic Coastal Plain. Bull. Amer. Paleont. 5 (24).

—— 1916. New Miocene fossils. Ibid. 5 (28).

—— 1917. The Murfreesboro stage of our east coast Miocene. Ibid. 5 (28).

OSBON, CLARENCE C. 1919. Peat in the Dismal Swamp, Virginia and North Carolina. U. S. Geol. Surv. Bull. 711: 41–59.

PARKER, J. M. 1949. Outlier near Raleigh, North Carolina. Jour. Elisha Mitchell Sci. Soc. 65: 195.

PETTY, JULIAN JAY. 1926. The origin and occurrence of fulgurites in the Atlantic Coastal Plain. Amer. Jour. Sci., ser. 5, 31: 188–201.

PRATT, JOSEPH H. 1941. History of geological investigations in North Carolina. Jour. Elisha Mitchell Sci. Soc. 57: 295–305.

POSTLEY, OLIVE. 1938. Oil and gas possibilities in the Atlantic Coastal Plain from New Jersey to Florida. Bull. Amer. Assn. Petr. Geol. 22: 799–815.

PROUTY, W. F. 1934. Carolina Bays. Jour. Elisha Mitchell Sci. Soc. 50: 59–60.

—— 1934a. Fossil whales of the North Carolina Miocene. Ibid. 50: 52.

—— 1934b. "Meteor Craters" of the Carolinas. Ibid. 50: 48.

—— 1935. "Carolina Bays" and elliptical lake basins. Jour. Geol. 43: 200–207.

—— 1936. Geology of the coastal plain of North Carolina. Jour. Amer. Water Works Assn. 28: 484–491.

—— 1936a. Further evidence in regard to the origin of "Carolina Bays" and Elliptical Lake Basins. *Proc. Geol. Soc. Amer. for 1935*: 96–97.

—— 1946. Two peneplanes under the coastal plain. *Jour. Elisha Mitchell Sci. Soc.* **62**: 139–140.

—— 1946a. Atlantic Coastal Plain floor and continental slope of North Carolina. *Bull. Amer. Assn. Petr. Geol.* **30**: 1917–1920.

—— 1949. Ellipticity of Carolina Bays. *Jour. Elisha Mitchell Sci. Soc.* **65**: 195–196.

PROUTY, W. F., and H. W. STRALEY. 1938. Further studies of "Carolina Bays." *Geol. Soc. Amer. Proc. for 1937*: 104–105.

RAISZ, ERWIN J. 1934. Rounded lakes and lagoons of the coastal plains of Massachusetts. *Jour. Geol.* **42**: 839–848.

RANDOLPH, E. E. 1942. Important North Carolina raw materials and manufacturing facilities available for war use. *Jour. Elisha Mitchell Sci. Soc.* **58**: 126–127.

RICHARDS, HORACE G. 1936. Some shells from the North Carolina "Banks." *Nautilus* **49**: 130–134.

—— 1936a. Fauna of the Pleistocene Pamlico formation of the southern Atlantic Coastal Plain. *Bull. Geol. Soc. Amer.* **47**: 1611–1656.

—— 1941. New mollusks from the Trent formation (Miocene) of North Carolina. *Ibid.* **52**: 1974.

—— 1943. Pliocene and Pleistocene fossils from the Santee-Cooper area, South Carolina. *Notulae Naturae, Acad. Nat. Sci. Phila.* 118.

—— 1943a. Studies on the geology and paleontology of the North Carolina Coastal Plain. *Amer. Philos. Soc. Year Book for 1942*: 110–119.

—— 1943b. Additions to the fauna of the Trent Marl of North Carolina. *Jour. Paleont.* **17**: 518–526.

—— 1943c. Pliocene and Pleistocene mollusks from the Santee-Cooper area, South Carolina. *Acad. Nat. Sci. Phila. Not. Nat.* 118.

—— 1945. Correlation of Atlantic Coastal Plain Cenozoic formations: a discussion. *Bull. Geol. Soc. Amer.* **56**: 401–408.

—— 1945a. Subsurface stratigraphy of Atlantic Coastal Plain between New Jersey and Georgia. *Bull. Amer. Assn. Petr. Geol.* **29**: 885–955.

—— 1945b. The subsurface stratigraphy of the Atlantic Coastal Plain. *Trans. N. Y. Acad. Sci.*, ser. 2, **8**: 1–4.

—— 1947. Invertebrate fossils from deep wells along the Atlantic Coastal Plain. *Jour. Paleont.* **21**: 23–37.

—— 1947a. Developments in Atlantic coastal states between New Jersey and North Carolina in 1946. *Bull. Geol. Soc. Amer.* **31**: 1106–1108.

—— 1947b. The Atlantic Coastal Plain, its geology and oil possibilities. *World Oil* 127 (3): 44 *et seq.*

—— 1948. Tertiary invertebrate fossils from newly discovered localities in North and South Carolina. Part 1. *Acad. Nat. Sci. Phila. Not. Nat.* 207.

—— 1948a. Studies on the subsurface geology and paleontology of the Atlantic Coastal Plain. *Proc. Acad. Nat. Sci. Phila.* **100**: 39–76.

—— 1948b. Developments in Atlantic coastal states between New Jersey and North Carolina in 1947. *Bull. Amer. Assn. Petrol. Geol.* **32**: 1077–1078.

—— 1949. Developments in Atlantic coastal states between New Jersey and North Carolina in 1948. *Bull. Amer. Assn. Petrol. Geol.* **33**: 1011.

—— 1949a. The occurrence of Triassic rocks in the subsurface of the Atlantic Coastal Plain. *Proc. Penna. Acad. Sci.* **23**: 45–48.

—— 1950. Developments in Atlantic Coastal States between New Jersey and North Carolina in 1949. *Bull. Amer. Assn. Petr. Geol.* **34**.

RICHARDS, HORACE G., AND ANNE HARBISON. 1942. Miocene invertebrate fauna of New Jersey. *Proc. Acad. Nat. Sci. Phila.* **94**: 167–250.

ROWLAND, H. I. 1936. The Atlantic and Gulf Coast Tertiary Pectinidae of the United States. Part 1. *Amer. Mid. Nat.* **17**: 471–490; Part 2. *Ibid.* 985–1017.

ROBERTS, JOSEPH K. 1932. The Lower York-James Peninsula. *Va. Geol. Surv. Bull.* 37.

RUDE, G. T. 1923. Shore changes at Cape Hatteras. *Annals Assn. Amer. Geog.* **12**: 87–95.

SCHUCHERT, CHARLES. 1943. Stratigraphy of the eastern and central United States. N. Y., Wiley.

SHELDON, PERAL G. 1917. The Atlantic Slope *Arcas. Paleontographica Americana* **1** (1).

SIPLE, GEORGE. 1946. Ground-water investigations in South Carolina. *Research, Planning and Development Board, Columbia, S. C. Bull.* 15.

SMITH, BURNETT. 1940. Notes on giant fasciolarias. *Paleontographica Americana* **2** (11).

SPANGLER, WALTER. 1950. Subsurface Geology of Atlantic Coastal Plain of North Carolina. *Bull. Amer. Assn. Petr. Geol.* **34**: 100–132.

SPANGLER, WALTER, and JAHN J. PETERSON. 1950. Geology of Atlantic Coastal Plain in New Jersey, Delaware, Maryland and Virginia. *Ibid.*, **34**: 1–99.

STENZEL, H. B. 1935. Nautiloids of the genus *Aturia* from the Eocene of Texas and Alabama. *Jour. Paleont.* **9**: 551–562.

STETSON, HENRY C. 1949. The sediments and stratigraphy of the east coast continental margin; Georges Bank to Norfolk Canyon. *Papers in Physical Oceanography and Meteorology, Mass. Inst. Tech. and Woods Hole Ocean. Inst.* **11** (2).

STEPHENSON, L. W. 1914. Cretaceous deposits of the eastern Gulf region and species of exogyra from the eastern Gulf region. *U. S. Geol. Surv. Prof. Paper* **81**.

—— 1914. A deep well at Charleston, South Carolina. *Ibid.* **90-H**.

—— 1926. Major features in the geology of the Atlantic and Gulf Coastal Plain. *Jour. Wash. Acad. Sci.* **16**: 460–480.

—— 1927. Additions to the Upper Cretaceous faunas of the Carolinas. *Proc. U. S. Nat. Mus.* 72, art. 10.

—— 1928. Structural features of the Atlantic and Gulf Coastal Plain. *Bull. Geol. Soc. Amer.* **39**: 887–900.

—— 1933. The zone of *Exogyra cancellata* traced 2500 miles. *Bull. Amer. Assn. Petrol. Geol.* **17**: 1351–1361.

—— 1936. Upper Cretaceous fossils from Georges Bank (including species from Banquereau, Nova Scotia). *Bull. Geol. Soc. Amer.* **47**: 367–412.

STEPHENSON, L. W., and M. J. RATHBUN. 1923. The Cretaceous formations of North Carolina. *N. C. Geol. and Econ. Surv.* 5.

STEPHENSON, L. W., C. W. COOKE, and JULIA GARDNER. 1939. The Atlantic and Gulf Coastal Plain. *In* Geology of North America 1. Geologie der Erde, Berlin.

STEPHENSON, L. W., F. B. KING, W. H. MONROE, and R. W. IMLAY. 1942. Correlation of the outcropping Cretaceous formations of the Atlantic and Gulf Coastal Plain and Trans-Pecos Texas. *Bull. Geol. Soc. Amer.* **53**: 435–448.

STOKES, A. P. DACHNOWSKI, and B. W. WELLS. 1929. The vegetation, stratigraphy and age of the "Open Land" peat area in Carteret County, North Carolina. *Jour. Wash. Acad. Sci.* **19**: 1–40.

STRALEY, H. W., and HORACE G. RICHARDS. 1949. The Atlantic Coastal Plain. *18th Internatl. Geol. Congr., London* (1948). Part 6.

STUCKEY, JASPER L. 1928. A Cretaceous sandstone quarry near Kinston, North Carolina. *Jour. Elisha Mitchell Sci. Soc.* **44**: 22–23.

—— 1949. North Carolina (Report of oil activities). *Interstate Oil Compact Commission Quart.* **8**: 27–29.

SWAIN, F. M. 1947. Two recent wells in Coastal Plain of North Carolina. *Bull. Amer. Assn. Petrol. Geol.* **31**: 2054–2060.

—— 1950. Mesozoic Ostracoda from subsurface of eastern North Carolina (abstract). Program of Chicago Meeting of Amer. Assn. Petr. Geol., 45.

TABER, STEPHEN. 1938. Geology of the South Carolina Coastal Plain in the vicinity of the Santee-Cooper project. *Bull. Geol. Soc. Amer.* **49**: 1963.

THOM, W. T., JR. 1937. Position, extent and structural makeup of Appalachia. *Bull. Geol. Soc. Amer.* **48**: 315–322.

TUCKER, H. I. 1931. Some new Tertiary Pectinidae. *Proc. Indiana Acad. Sci.* **40**: 243–245.

—— 1934. Some Atlantic Coast Tertiary Pectinidae. *Amer. Mid. Nat.* **15**: 612–618.

TUCKER-ROWLAND, H. I. 1928. The Atlantic and Gulf Coast Tertiary Pectinidae of the United States—Part 3. *Mus. Roy. Hist. Nat. Belgique,* 2 ser., 13.

TYLER, S. A. 1934. A study of sediments from the North Carolina and Florida coasts. *Jour. Sed. Petrol.* **4**: 3–11.

VAUGHAN, T. W. 1918. Correlation of the Tertiary geologic formations of the southeastern United States, Central America and the West Indies. *Jour. Wash. Acad. Sci.* **8**: 268–276.

VEATCH, OTTO, AND L. W. STEPHENSON. 1911. Preliminary report on the geology of the Coastal Plain of Georgia. *Geol. Surv. of Ga. Bull.* **26**.

WATSON, F., JR. 1936. Meteor craters. *Pop. Astron.* **44**: 2–17.

WELLS, B. W. 1940. Preliminary survey of the eastern Dare County peat. *Jour. Elisha Mitchell Sci. Soc.* **56**: 219.

—— 1942. Ecological problems of the southeastern United States Coastal Plain. *Bot. Rev.* **8**: 533–561.

—— 1943. Blythe Bay: a record of changing ocean level. *Jour. Elisha Mitchell Sci. Soc.* **59**: 118–119.

—— 1949. Origin of the Carolina Bays: Evidence from some peat profiles. *Ibid.* **65**: 185.

WENTWORTH, C. K. 1930. Sand and gravel resources of the Coastal Plain of Virginia. *Va. Geol. Surv. Bull.* **32**.

WYLIE, C. C. 1933. Iron meteorites and the Carolina "Bays." *Pop. Astron.* **41**: 410–412.

—— 1933a. On the formation of meteoric craters. *Ibid.* **41**: 211–214.

BIBLIOGRAPHY B

List of additional references cited in the text of this report including some published prior to 1912 as well as more recent works dealing largely with regions outside of North Carolina Coastal Plain.

BAYLEY, W. S. 1925. The kaolins of North Carolina. *N. C. Geol. & Econ. Surv. Bull.* **29**.

BROWN, CARL B. 1932. A new Triassic area in North Carolina. *Amer. Jour. Sci.,* ser. 5, **23**: 525–528.

CAMPBELL, M. R. 1931. Alluvial fan of the Potomac River. *Bull. Geol. Soc. Amer.* **42**: 825–852.

CARPENTER, F. B. 1894. Marls and phosphates of North Carolina. *N. C. Agric. Exp. Station Bull.* **110**.

CLARK, W. B. 1915. The Brandywine formation of the Middle Atlantic Coastal Plain. *Amer. Jour. Sci.* **40**: 499–506.

CLARK, W. B., and B. L. MILLER. 1906. The geology of the Virginia Coastal Plain. *Va. Geol. Surv. Bull.* **2**: 12–24.

CONRAD, T. A. 1872. Descriptions of a new Recent species of *Glycymeris* from Beaufort, North Carolina, and of Miocene shells of North Carolina. *Proc. Acad. Nat. Sci. Phila.* **24**: 216–217.

DALL, W. H. 1892. Contributions to the Tertiary fauna of Florida (part 2). *Trans. Wagner Free Inst. Sci.* **3** (2).

—— 1895. Diagnoses of new Tertiary fossils from the southern United States. *Proc. U. S. Natl. Mus.* **18**: 21–46.

HILGARD, E. W. 1891. Orange sand, Lagrange and Appomattox. *Amer. Geol.* **8**: 129–131.

KERR, W. C. 1875. Report on the geological survey of North Carolina 1. Raleigh, N. C.

NUTTING, P. G. 1933. The bleaching clays. *U. S. Geol. Surv. Circular* **3**.

RICHARDS, HORACE G. 1931. The occurrence of old meadow sod under the New Jersey beaches. *Science* **73**: 673–674.

—— 1934. Is the coast of New Jersey sinking? *Nature Mag.* **24**: 225–226.

—— 1938. Marine Pleistocene of Florida. *Bull. Geol. Soc. Amer.* **29**: 1267–1296.

—— 1939. Marine Pleistocene of the Gulf Coastal Plain: Alabama, Mississippi and Louisiana. *Ibid.* **50**: 1885–1898.

—— 1939. Marine Pleistocene of Texas. *Bull. Geol. Soc. Amer.* **50**: 1885–1898.

—— 1943. Fauna of the Raritan formation of New Jersey. *Proc. Acad. Nat. Sci. Phila.* **95**: 15–32.

RUFFIN, EDMUND. 1843. Report of the commencement and progress of the agricultural survey of South Carolina for 1843. Columbia, S. C.

SHATTUCK, G. B. 1901. The Pleistocene problem of the North Atlantic Coastal Plain. *Johns Hopkins Univ. Circ.* **20**: 74.

—— 1902. The Miocene formation of Maryland. *Science* **15**: 906–907.

SLOAN, EARLE. 1904. A preliminary report on the clays of South Carolina. *S. C. Geol. Surv. Set. 4, Bull.* **1**.

—— 1907. Handbook of South Carolina. S. C. Dept. Agric., Commerce and Immigration.

SMITH, E. A., and L. C. JOHNSON. 1887. Tertiary and Cretaceous strata of the Tuscaloosa, Tombigbee and Alabama Rivers. *U. S. Geol. Surv. Bull.* **43**.

STEPHENSON, L. W. 1907. Some facts relating to the Mesozoic deposits of North Carolina. *Johns Hopkins Univ. Circ.* n. s. 7.

FIG. 12. Sand pit in Tuscaloosa formation, Sanford, Lee County, N. C.

FIG. 15. Pleistocene Coharie formation overlying the Upper Cretaceous Black Creek formation near Smithfield, Johnson County, N. C.

FIG. 13. Exposure of Tuscaloosa formation along Seaboard Air Line Railroad, Southern Pines, Moore County, N. C.

FIG. 16. Detail of same locality.

FIG. 14. Tuscaloosa formation at Spout Springs, Harnett County, N. C.

FIG. 17. Continental phase of Black Creek formation in Fort Bragg Reservation near Fayetteville, Cumberland County, N. C.

FIG. 18. Exposures of Middle Eocene quartzite near Clayton, Johnson County, N. C.

FIG. 21. Quarry at Castle Hayne, New Hanover County, N. C.

FIG. 19. Exposure of Middle Eocene (?) sand in pit at Lillington, Harnett County, N. C.

FIG. 22. Castle Hayne marl dug from pit at Cedar Fork Swamp, 5 miles east of Beulaville, Duplin County, N. C.

FIG. 20. Pliocene Waccamaw formation overlying the Cretaceous Peedee on Cape Fear River at Neils Eddy Landing, Columbus County, N. C.

FIG. 23. Ledge of Castle Hayne limestone at Chinquipin on Northeast Cape Fear River, Duplin County, N. C.

66

Fig. 24. Pit showing specimens of *Ostrea georgiana* in Castle Hayne formation at Pollocksville, Jones County, N. C.

Fig. 25. Quarry of Superior Stone Company at Belgrade, Onslow County, N. C.

Fig. 26. Exposure of Yorktown Miocene formation at Tar Landing Ferry near Harrellsville, Hertford County, N. C.

Fig. 27. Simmons Marl Pit, 3 miles east Pollocksville, Jones County, N. C.

Fig. 28. Cemetery wall at New Bern, Craven County, N. C., built of Trent limestone.

Fig. 29. Exposure of Trent limestone on Neuse River at Spring Garden, 3½ miles northeast of Jasper, Craven County, N. C.

67

FIG. 36. Watson's Mill (Worrell's Mill) on Kirby Creek, 2½ miles northwest of Murfreesboro, Northampton County, N. C.

FIG. 39. Exposure of Waccamaw Pliocene formation near Brown's Creek, 1½ miles south of Elizabethtown, Bladen County, N. C.

FIG. 37. Lake Waccamaw, Columbus County, N. C. Pliocene Waccamaw formation overlying the Duplin Miocene.

FIG. 40. Croatan formation on Neuse River near James City, Craven County, N. C.

FIG. 38. Natural well, near Magnolia, Duplin County, N. C.

FIG. 41. Santee-Cooper Diversion Canal, Berkeley County. S. C.

Fig. 42. Bonsall Gravel Pit, Lilesville, Anson County, N. C.

Fig. 45. "High Level Gravels" near Stanhope, Nash County, N. C.

Fig. 43. Pliocene (?) gravel in pit of Cape Fear Gravel Company, 2 miles northwest of Lillington, Harnet County, N. C.

Fig. 46. Pleistocene Pamlico formation along Neuse River, 10 miles south of New Bern, Craven County, N. C.

Fig. 44. "High Level Gravels" near Stanhope, Nash County, N. C.

Fig. 47. Pleistocene cypress stumps along Neuse River, 10 miles south of New Bern, Craven County, N. C.

70

Fig. 48. Pleistocene coquina south of Carolina Beach, New Hanover County, N. C.

Fig. 51. Pleistocene fossils dredged from Intracoastal Canal, Hyde County, N. C.

Fig. 49. Chowan formation near Mount Gould, Bertie County, N. C.

Fig. 52. Intracoastal Canal near Myrtle Beach, Horry County, S. C.

Fig. 50. Pamlico formation on north bank of Neuse River at Bennett Plantation, Pamlico County, N. C.

Fig. 53. Tree stumps on beach near Kitty Hawk, Dare County, N. C.

71

FIG. 54. Setting up the rig, Carolina Petroleum Company well near Merrimon, Carteret County, N. C. (July, 1946).

FIG. 55. "High Level Gravels" on Tuscaloosa clay near Lilesville, Anson County, N. C.

FIG. 56. Alligator Lake, Hyde County, N. C.

FIG. 57. Esso No. 2 well in Pamlico Sound, Dare County, N. C. (February, 1947).

FIG. 58. Esso No. 1 well near Hatteras Lighthouse, Dare County, N. C. (July, 1946).

FIG. 59. Granite outcrop, Fountain, Pitt County, N. C.

FIG. 60. Granite outcrop near Smithfield, Johnston County, N. C.

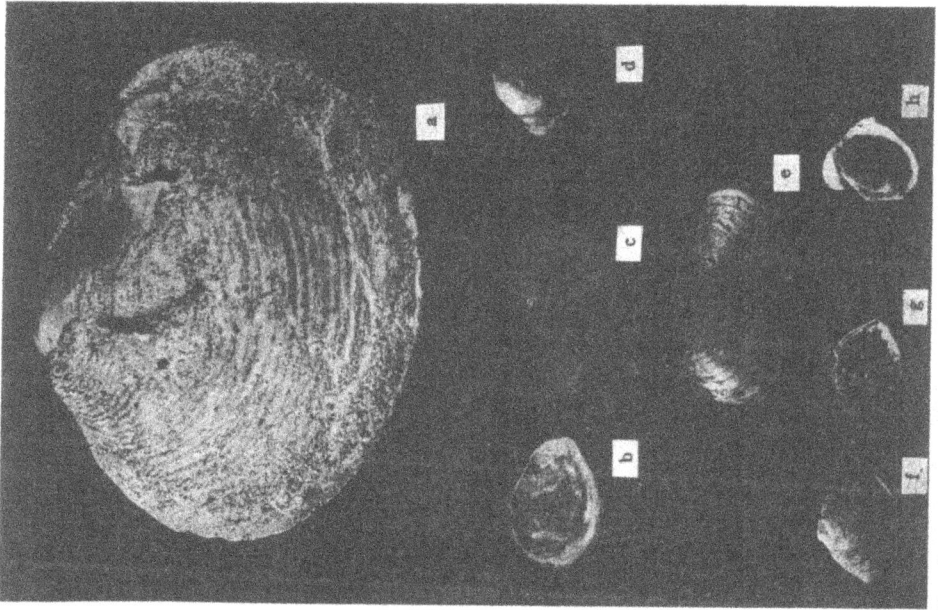

FIG. 61. a, Crassatellites wilcoxi Brown and Pilsbry, Eocene, Wilmington, N. C. b, c, C. pteropsis Conrad, Upper Cret., Snow Hill, N. C. d, h. Exogyra woolmani Richards, Upper Cret., Norfolk, Va., 484 feet. e. Linuaria carolinensis Conrad, Upper Cret., Snow Hill, N. C. f, g. Lucina parva Stephenson, Upper Cret., Kinston, N. C., 60 feet. Figures 61–76 show fossils slightly smaller than natural size.

FIG. 62. a. Aturia alabamensis Morton, Eocene, Wilmington, N. C. b. Exogyra costata Say, Upper Cret., Myrtle Beach, S. C. c, d. Pecten cushmani Kellum, Lower Miocene, Quinnerly, N. C. e. Pecten cookei Kellum, Lower Miocene, near Magnolia, N. C. f. Cardium penderense Stephenson, Upper Cret., Wilmington, N. C. g. Gryphaeostrea vomer Morton, Eocene, Williamston, N. C., 110 feet.

74

FIG. 30. Tar Landing Ferry near Harrellsville, Hertford County, N. C.

FIG. 33. Exposure of Yorktown fossils at Colerain, on Chowan River, Bertie County, N. C.

FIG. 31. Exposure of Yorktown formation at Black Rock on Chowan River 1½ miles above Eden House Point, Bertie County, N. C.

FIG. 34. Exposure of Yorktown fossils at Old Sparta, Edgecombe County, N. C.

FIG. 32. Palmyra Bluff on south bank Roanoke River, Halifax County, N. C.

FIG. 35. Marl pit near Maury, Greene County, N. C.

68

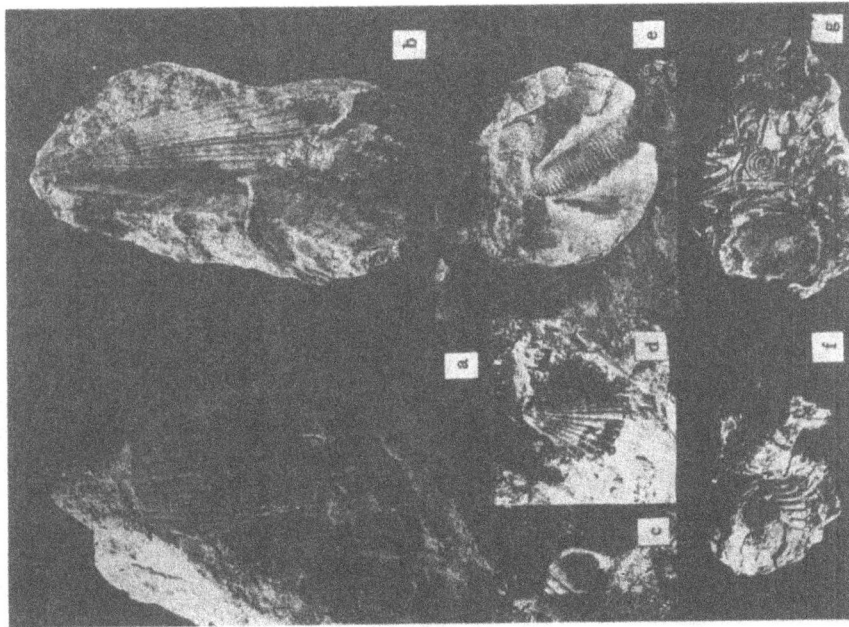

Fig. 63. *a, b, Pinna harnetti* Richards, Eocene, Spout Springs, N. C. *c. Amaurellina* sp., Eocene, Crabtree State Park, N. C. *d. Pecten membranosus* Morton, Eocene, Spout Springs, N. C. *e. Schizaster armiger* Clark, Eocene, Spout Springs, N. C. *f. Venericardia planicosta* Lamarck var. Eocene, Clayton, N. C. *g. Spirulaea rotula* Morton, Eocene, Clayton, N. C.

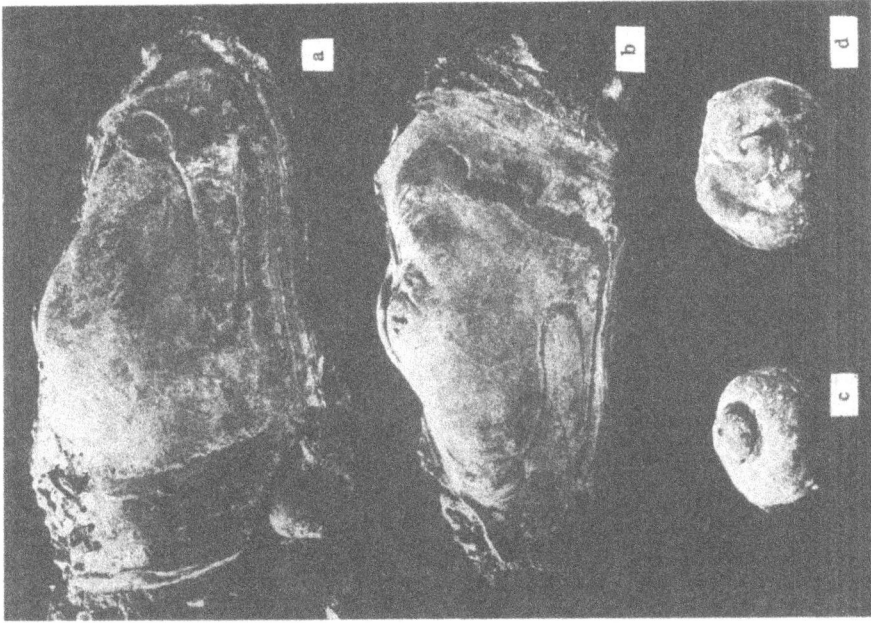

Fig. 64. *a, b, Panope intermedia* Richards, Lower Miocene, Belgrade, N. C. *c, d. Calyptraea aperta* Solander, Lower Miocene, Belgrade, N. C.

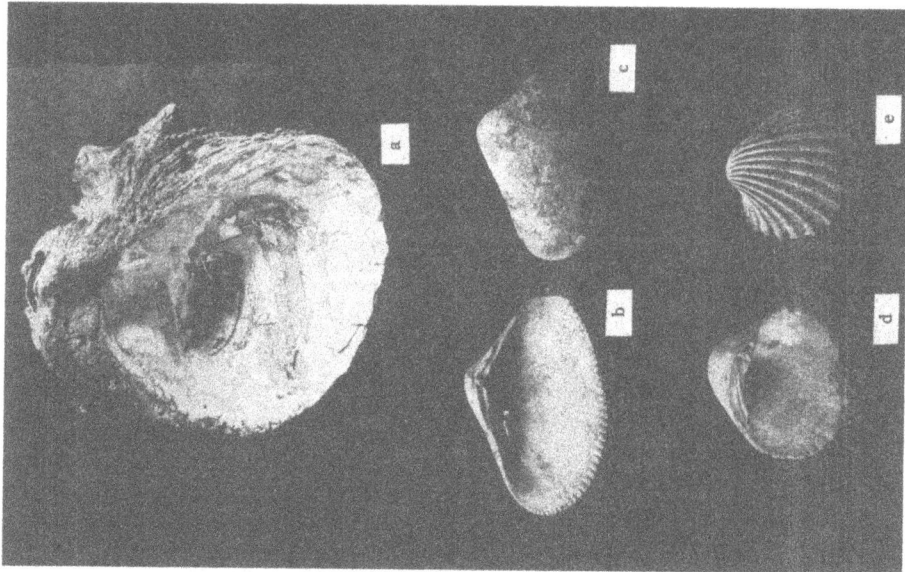

Fig. 66. *a. Ostrea georgiana* Conrad var. *b, c. Donax idonens* Conrad. *d, e. Venericardia nodifera* Kellum. (All Lower Miocene, from Silverdale, N. C.)

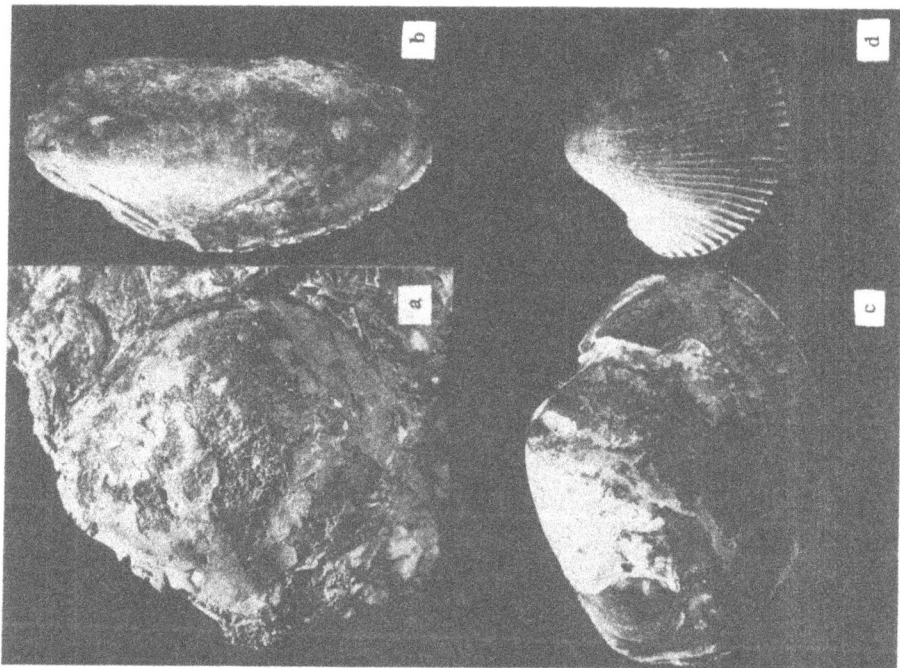

Fig. 65. *a. Ostrea* cf. *thirsæ* Gabb, Eocene, Clayton, N. C. *b. Modiolus stuckeyi* Richards, Lower Miocene, Belgrade, N. C. *c. Venus gardnerae* Kellum. Lower Miocene, Belgrade, N. C. *d. Cardium belgradensis* Richards, Lower Miocene, Belgrade, N. C.

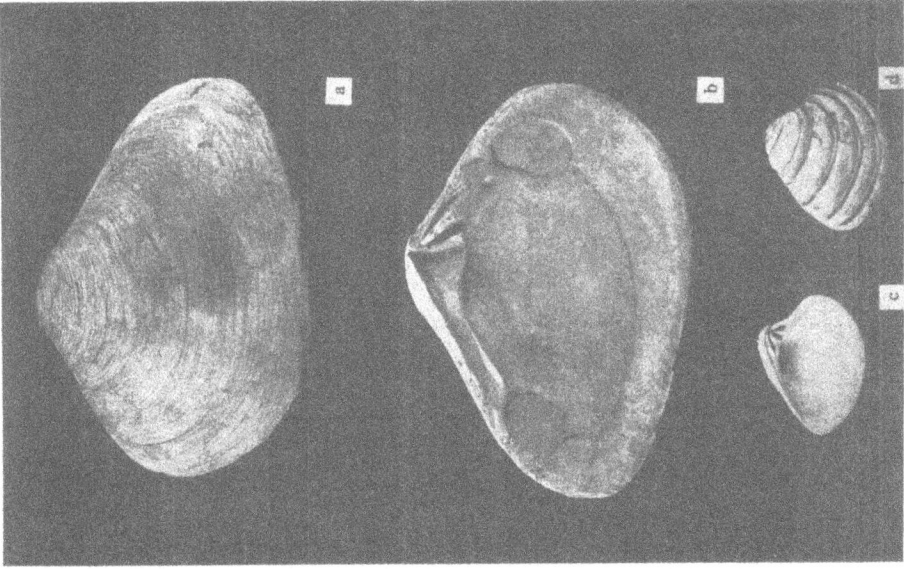

Fig. 68. a, b. Crassatellites undulatus Say, Miocene, Williamston, N. C. c, d. Chione latilirata Conrad, Pliocene, Acme, N. C.

Fig. 67. a. Pecten trentensis Harris. b, c. Conus postalveatus Kellum. d. Turritella fuerta Kellum. e. Plicatula densata Conrad. f. Busycon spiniger onslowensis Kellum. g. Potamides silverdalensis Richards. h. Cardium midyetti Richards. i. Calyptraea trochiformis Conrad. j. Murex davisi Richards. k. Ostrea georgiana Conrad var. (All Lower Miocene; a, from Pollocksville, b-k, from Silverdale, N. C.)

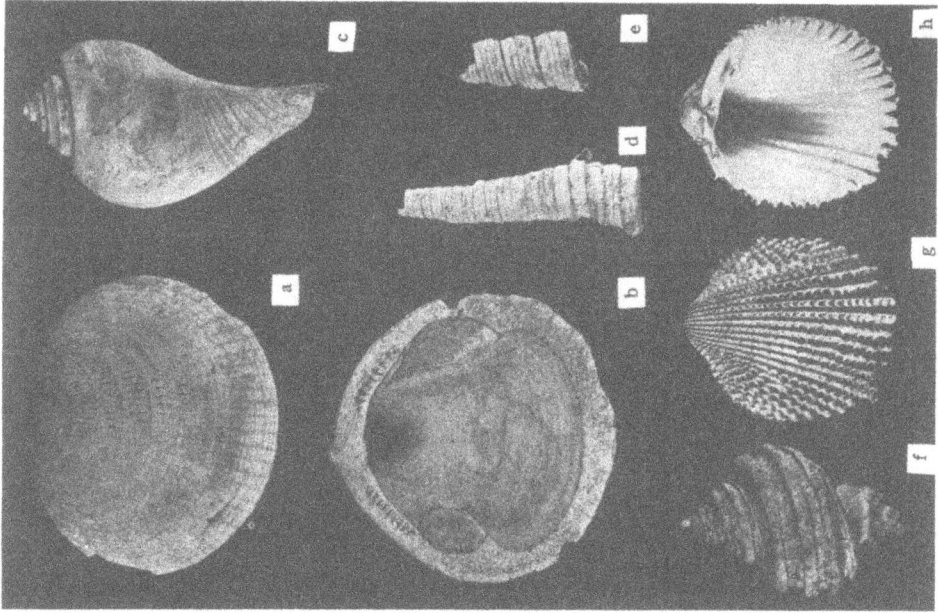

Fig. 70. *a, b. Glycymeris subovata* Say, Miocene, Maury, N. C. *c. Fulgur excavatum* Conrad, Miocene, Natural Well, N. C. *d. Vermetus sculpturata* H. C. Lea, Miocene, Tar Heel, N. C. *e. Turritella variabilis* Conrad, Miocene, Fort Barnwell, N. C. *f. Ecphora quadricostata* Say, Miocene, Maury, N. C. *g, h. Cardium isocardia* Linné, Pliocene, Neils Eddy Landing, N. C.

Fig. 69. *a. Pecten jeffersonius* Say, Miocene, North Carolina. *b, c. Plicatula marginata* Say, Miocene, Maury, N. C. *d, e. Astarte concentrica bella* Conrad, Miocene, Fort Barnwell, N. C. *f, g. Astarte obruta* Conrad, Miocene, Watsons Mill, N. C. *h. Arca carolinensis* Wagner, Miocene, Tar Ferry, N. C.

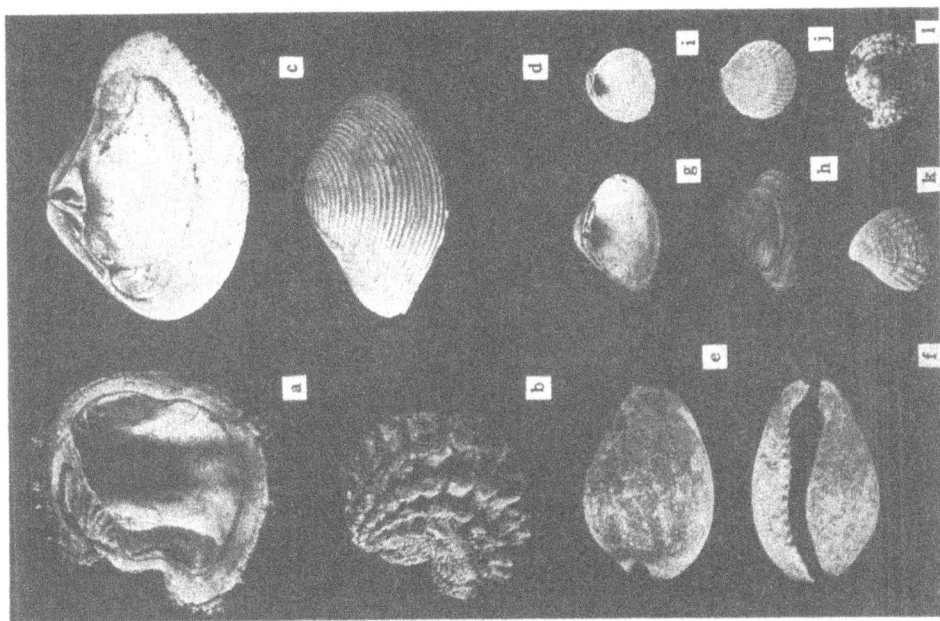

FIG. 72. a, b. *Echinochama arcinella* Linné, Pliocene, Acme, N. Y. c, d. *Crassatellites gibbersii* T. & H., Pliocene, Acme, N. C. e, f. *Cypraea carolinensis* Conrad, Pliocene, Tar Heel, N. C. g, h. *Mulinia congesta* Conrad, Miocene, Maury, N. C. i, j. *Venericardia granulata* Conrad, Miocene, Maury, N. C. k. *Glans perplanus* Conrad, Miocene, Bacons Castle, Va., 35 feet. l. *Calliostoma ruffini* H. C. Lea, Miocene, Bacons Castle, Va., 35 feet.

FIG. 71. a. *Mitra carolinensis* Dall, Pliocene, Tar Heel, N. C. b. *Pecten eboreus* Conrad, Pliocene. c. *Fulgur maximun* Conrad. d. *Fasciolaria elegans* Emmons. (All Pliocene, from Tar Heel, N. C.)

79

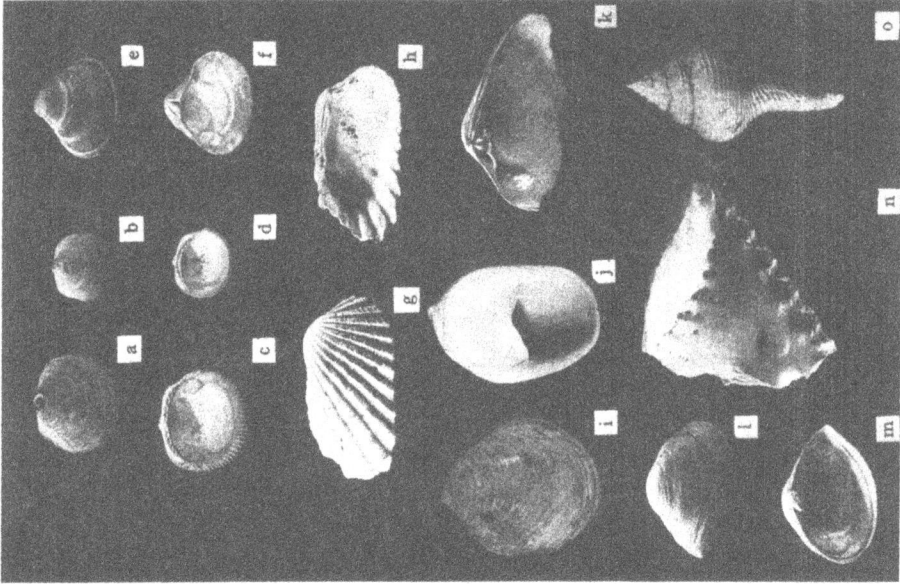

FIG. 74. a-d. *Glycymeris boguensis* Richards and Harbison, Miocene, Bogue, N. C., 130 feet. *e, f. Astarte mundorffi* R. & H., Miocene, Bogue, N. C., 130 feet. *g, h. Cardita arata* Conrad, Pliocene, Tar Heel, N. C. *i, j. Crepidula fornicata* Say, Pliocene, Tar Heel, N. C. *k. Macrocallista* sp., Miocene, Atlantic, N. C., 140 feet. *l, m. Corbula inaequalis mansfieldi* Richards, Pliocene, Harvey Neck, N. C. *n. Melina maxillata* Deshayes, Miocene, Bogue, N. C., 130 feet. *o. Fusus* sp., Pliocene, Tar Heel, N. C.

FIG. 73. *a. Dentalium* sp., Miocene, Watsons Mill, N. C. *b. Terebra carolinensis* Conrad, Miocene, Elizabethtown, N. C. *c. Teredo fistula* Lea, Miocene, Watsons Mill, N. C. *d. Vermetus graniferus* Say, Pliocene, Tar Heel, N. C. *e, f. Cypraeolina dacria* Dall, Miocene, Edenton, N. C., 72 feet. *g, h. Crucibulum constrictum* Conrad, Miocene, Maury, N. C. *i. Conus adversarius* Conrad, Pliocene, Tar Heel, N. C. *j, k. Oliva sayana* Ravenel, Pliocene, Acme, N. C. *l, m. Rangia* sp., Pliocene, Tar Heel, N. C. *n. Spissula marylandica* Dall, Miocene, Edenton, N. C., 60 feet. *o. Sphenia dubia* H. C. Lea, Miocene,

80

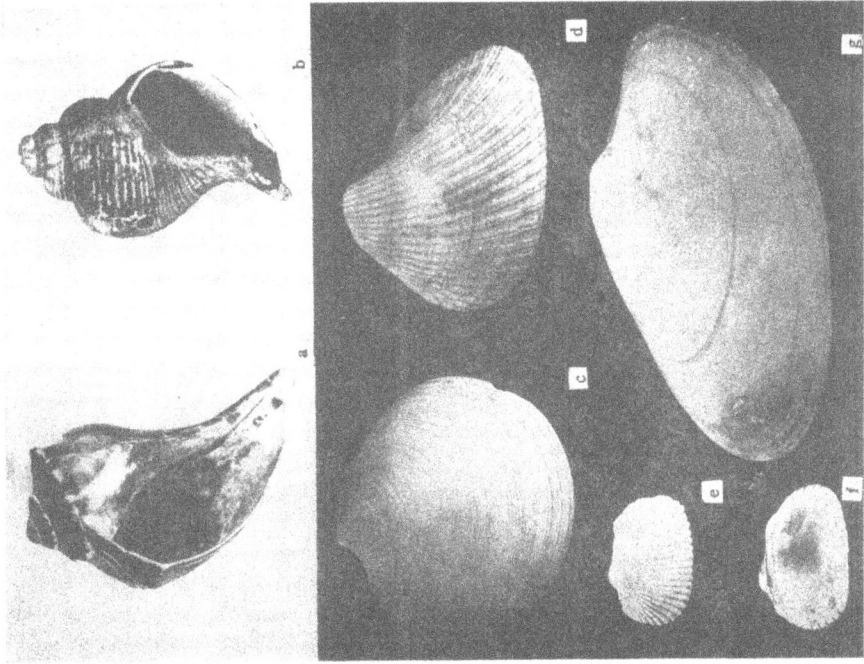

Fig. 76. a. *Fulgur perversum* Linné. b. *Neptunea stonei* Pilsbry. c. *Dosinia discus* Reeve. d. *Arca ponderosa* Say. e, f. *Arca transversa* Say. g. *Macrocallista nimbosa* Solander. (All Pleistocene; a, b, from New Jersey; c-g, from Hyde County, N. C.)

Fig. 75. a. *Cardium robustum* Solander. b. *Arca campechiensis* Gmelin. c. *Pecten gibbus* Linné. d, o, p. *Terebra dislocata* Say. e. *Urosalpinx cinerea* Say. f. *Chione cancellata* Linné. g. *Polinices duplicata* Say. h, i. *Olivella mutica* Say. j, n. *Tellina lintea* Conrad. k. *Astrangia danae* Agassiz. l. *Cantharus cancellaria* Conrad. m. *Rangia cuneata* Gray. (All Pleistocene, from excavations for Inland Waterway, Hyde County, N. C.)

81

INDEX

a. LOCALITIES

b. FIGURED SPECIES